Leatherhead Food International

INGREDIENTS HANDBOOK

PREBIOTICS AND PROBIOTICS
(2nd Edition)

Edited by
Shelly Jardine

WILEY-BLACKWELL

This edition first published 2009 by
Leatherhead Publishing
a division of
Leatherhead International Ltd
Randalls Road, Leatherhead, Surrey KT22 7RY, UK
URL: http://www.leatherheadfood.com

and

Blackwell Publishing Ltd

Blackwell Publishing was acquired by John Wiley & Sons in February 2007.
Blackwell's publishing programme has been merged with Wiley's global Scientific,
Technical, and Medical business to form Wiley-Blackwell.

Registered office
John Wiley & Sons Ltd, The Atrium, Southern Gate, Chichester, West Sussex, PO19
8SQ, United Kingdom

Editorial offices
9600 Garsington Road, Oxford, OX4 2DQ, United Kingdom
2121 State Avenue, Ames, Iowa 50014-8300, USA

www.wiley.com/wiley-blackwell

ISBN: 978-1905224-52-4

A catalogue record of this book is available from the British Library

Typeset by Alison Turner
 Index by Ann Pernet
Printed and bound by the MPG Books Group in the UK

Books are to be returned on or before
the last date below.

CONTENTS

CONTRIBUTORS

Hélène Alexiou
BENEO- Ofrati
Aandorenstraat 1,
3300 Tienen
Belgium

Fiona Angus
Head of Nutrition
Leatherhead Food International
Randalls Road
Leatherhead
Surrey, KT22 7RY
UK

Dr Anne Franck
BENEO-Orafti
Aandorenstraat 1,
3300 Tienen
Belgium

Shelly Jardine
Market and Technical Services
Leatherhead Food International
Randalls Road
Leatherhead
Surrey, KT22 7RY
UK

Sophia Johansson
Regulatory Services
Leatherhead Food International
Randalls Road
Leatherhead
Surrey, KT22 7RY
UK

Arjen Nauta
FrieslandCampina DOMO
Hanzeplein 25
8017 JD Zwolle,The Netherlands

Margaret O'Connell
Chr. Hansen Ltd.,
2 Tealgate,
Hungerford, RG17 0YT
UK

Pauline Quierzy
Solvay SA
25 Rue de Clichy
F-75442 Paris Cedex 09
France

Dr Robert A Rastall
Department of Food Biosciences
School of Chemistry, Food Bioscince
and Pharmacy
University of Reading
Reading G81 2LG
UK

Dr Maria Saarela
VTT Technical Research Centre of
Finland
PO Box 1000, FIN-02044 VTT
Finland

H.C Schoterman
FrieslandCampina DOMO
Hanzeplein 25
8017 JD Zwolle,The Netherlands

Dimitrios K Tzimorotas
Department of Food Biosciences
School of Chemistry, Food Bioscince
and Pharmacy
University of Reading
Reading G81 2LG
UK

FOREWORD

The Ingredients Handbook series, published by Leatherhead Food International and Blackwell Publishing Ltd, constitutes a comprehensive source of information on food additives and ingredients including colours, sweeteners, fat replacers and pre- and probiotics. Each title in the series provides a concise overview of the general and physiological properties associated with the specific ingredient or additive, as well as detailed information on applications. In addition, all handbooks contain a review of the related legislation in the UK, mainland Europe and USA, and a detailed list of suppliers.

All chapters are written by experts from within the industry who manufacture or use these products on a daily basis, and edited by Leatherhead Food International's team of experienced technical advisors. These handbooks are an invaluable and authoritative source of reference for all food technologists, and other food-industry professionals. The first handbook on pre- and probiotics published in 2000 has been completely revised and updated in this Second Edition.

Acknowledgements

My thanks to all the authors who have contributed their time, expertise and advice to this publication, and also to their employers for supporting them. I would also like to acknowledge Catherine Hill, my manager at Leatherhead Food International, who has been an exceptional guide throughout, Victoria Emerton, team leader of the scientific and technical team at Leatherhead Food International, for her excellent advice and patience with editing this book; Fiona Angus for her technical advice; friends and colleagues for their patience; and my husband Chris for all the extra support at home while I was surrounded by paper.

Shelly Jardine
Leatherhead Food International

INTRODUCTION

Fiona Angus
Head of Nutrition
Leatherhead Food International
Randalls Road
Leatherhead
Surrey, KT22 7RY
UK

Probiotics and prebiotics have really been a success story within the global functional foods market. Gut health products have been long established in the Japanese food market, but they have recently shown great popularity in Europe and increasingly they are now making their presence felt in the United States, although this market has been harder to penetrate and it is only just becoming established. If a broad definition of the functional foods market is taken, then the global functional foods market is now estimated to be well in excess of US$40 billion. Gut health products are thought to account for over 40% of the global functional foods market and products containing prebiotics and probiotics dominate the sector.

Probiotics and prebiotics are both used to improve the microbial population of the gastrointestinal tract through dietary means. Probiotics are "live organisms that when ingested in adequate amounts confer a benefit to the host" (1), and prebiotics are defined as "non digestible food ingredients that benefit the host by selectively stimulating the growth and/or activity of one or more of a limited number of bacteria in the colon and thus improve health" (2).

This handbook is the Second Edition of the Probiotic and Prebiotics Handbook, first published by Leatherhead Food International in 2000. Much has happened in the field in nine years, most notably a growing body of science to support the health benefits of prebiotics and probiotics; the rapid growth of some probiotic food and drink products into established health brands; the application of probiotics in a range of end use applications including shelf stable products for the first time; the broadening application of prebiotics in a variety of food and drink products; and growing interest in new ingredients with possible prebiotic potential.

In the Second Edition of the handbook, all of the chapters have been fully updated to take account of changing regulations, emerging science and growing application experience. The section on miscellaneous probiotics has been removed and replaced by synbiotics - an application area showing great growth potential in the food industry.

The Regulatory section has also been updated to reflect legislation changes around the world. There have been major changes in the regulatory framework surrounding nutrition and health claims in Europe in the last nine years. Regulation 1924/2006, finally adopted in Europe in 2007, has radically changed the way in which nutrition and health claims are regulated and it is impacting on the claims that can be made about pre- and probiotics, as well as other functional food ingredients. Over the coming years, there will be significant changes in terms of the types of health claims that are made on packaging of prebiotics, probiotics and synbiotics.

Leatherhead Food International's Prebiotics and Probiotics Handbook is designed primarily to help product developers identify suitable pre- and probiotics for use in their products, providing a useful summary of the key physical properties of the ingredients and suitable applications. In addition, an update is provided as to the science behind their physiological and nutritional properties. There is also a useful suppliers section giving companies who can supply the ingredients. The book will prove an invaluable reference tool to anyone looking for a current picture of this important ingredient sector.

References

1. Food and Agriculture Organisation. World Health Organisation. Guidelines for the Evaluation of Probiotics in Food. Guidelines, Ontario, Canada, May 2002. Ed. Food and Agriculture Organisation, World Health Organisation Geneva, WHO, 2002.

2. Gibson G.R, Roberfroid M.B. Dietary modulation of the human colonic microbiota: introducing the concept of prebiotics. *Journal of Nutrition*, 1995, 125, 1401-12.

1. PREBIOTICS

1.1 INULIN AND OLIGOFRUCTOSE

Anne Franck, Ph.D. and Hélène Alexiou
BENEO-Orafti
Aandorenstraat 1
3300 Tienen
Belgium

1.1.1   Description

Inulin and oligofructose are ß(2-1) fructans, and are widely found in nature. Fructans are, after starch, the most abundant non-structural natural polysaccharides, and are present in a wide variety of plants and in some bacteria and fungi. Plants containing inulin (and oligofructose) primarily belong to either the Liliales, e.g. leek, onion, garlic and asparagus, or the Compositae, such as Jerusalem artichoke, dahlia, yacon and chicory, as shown in Table 1.1.I (1). Inulin functions as a carbohydrate reserve and enables plants to survive in the winter period.

TABLE 1.1.I
Inulin and oligofructose content in plants used for human nutrition (1)

Source	Edible part	Inulin content (% on fresh weight)	Oligofructose content (% on fresh weight)
Onion	Bulb	2-6	2-6
Jerusalem artichoke	Tuber	16-20	10-15
Chicory	Root	15-20	5-10
Leek	Bulb	3-10	2-5
Garlic	Bulb	9-16	3-6
Artichoke	Leaves-heart	3-10	<1
Banana	Fruit	0.3-0.7	0.3-0.7
Rye	Cereal	0.5-1.0	0.5-1.0
Barley	Cereal	0.5-1.5	0.5-1.5
Dandelion	Leaves	12-15	Not available
Burdock	Root	3.5-4.0	Not available
Camas	Bulb	12-22	Not available
Murnong	Root	8-13	Not available
Yacon	Root	3-19	3-19
Salsify	Root	4-11	4-11
Wheat	Cereal	1-4	1-4

Inulin and oligofructose have always been part of the normal human diet. Based on the consumption data of several plant foodstuffs, their average daily intake has been estimated at about 2-10 g per inhabitant in Europe, and 1-4 g in the United States (1). During the early nineties, several attempts were made to isolate and purify inulin and oligofructose for use as dietary supplements. Nowadays, they are used in a pure form as ingredients in many food products.

Given their high inulin content (> 15%), Jerusalem artichoke, dahlia and chicory were initially considered for industrial production in our temperate regions, but for several reasons chicory (*Cichorium intybus*) is nearly exclusively processed (2). The roots of chicory, which are also used in different countries for the production of a coffee substitute (after roasting), look like small oblong sugar beets. Their inulin content is high (more than 70% dry weight) and fairly constant from year to year.

The production of inulin was first described in 1920 by Schöne, who reported pilot tests, and in 1927 by Belval, who mentioned industrial trials (3,4). In 1931, a patent application describing an improvement in the extraction process was filed by Raffinerie Tirlemontoise (Tienen, Belgium). Inulin, which is sold as an industrial food ingredient, has been manufactured for nearly 20 years by BENEO-Orafti (Belgium), Cosucra (Belgium) and Sensus (The Netherlands). The production, represented in Figure 1.1.1, involves the extraction of the naturally occurring inulin from chicory roots in a process very similar to the extraction of sucrose from sugar beets (diffusion in hot water), followed by refining using technologies from the sugar and starch industries (e.g. ion exchangers), and then evaporation and spray-drying (2).

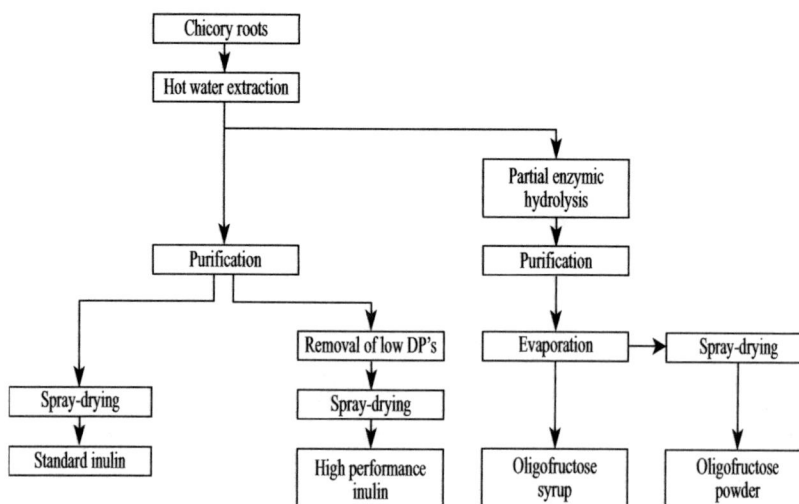

DP = degree of polymerisation

Fig. 1.1.1. Production process of chicory inulin and oligofructose

Oligofructose (or fructo-oligosaccharides) is produced using two different manufacturing techniques which deliver slightly different end products. BENEO-Orafti (Belgium) produce oligofructose by partial enzymic hydrolysis of chicory inulin (using a specific endo-inulinase), possibly followed by spray-drying (Figure 1.1.1), whereas Syral (France) synthesise it from sucrose using fructosyl-transferase (5). The latter product is sometimes also referred to as short-chain fructo-oligosaccharides.

1.1.2 General properties

1.1.2.1 Structure

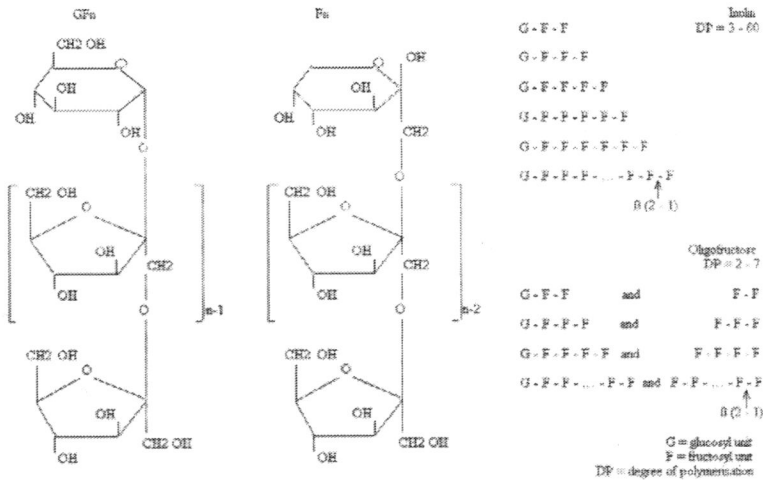

Fig. 1.1.2. Chemical structure of inulin and oligofructose

Inulin, as shown in Figure 1.1.2, is a polydisperse ß(2-1) fructan composed of a mixture of oligo- and polysaccharides that are almost all linear chains of fructose having the structure GF_n (with G=glucosyl unit, F=fructosyl unit and n=number of fructosyl units linked to one another) (6). In native chicory inulin (standard) the number of fructose units linked together ranges from 2 to about 65, with an average degree of polymerisation (DP) of the order of 10-12. Long-chain inulin (or high performance inulin) from which the lower DP-fraction has been

5

physically removed and having an average DP of about 25 is also available, and is used for functions such as texture improvement and fat replacement.

Oligofructose obtained from inulin contains GF_n and F_n chains with n ranging from 2 to 8, whereas fructo-oligosaccharides produced from sucrose only have GF_n forms with n between 2 and 4 (5, 6).

A specific combination of long-chain inulin and oligofructose (1:1), known as Synergy1, has also been developed in order to offer enhanced nutritional benefits.

Chicory inulin is available as white, odourless powders and oligofructose as powders and colourless viscous syrups (about 75% dry substance), all with a high purity and a well-known chemical composition. Their physico-chemical properties are summarised in Table 1.1.II.

TABLE 1.1.II
Physico-chemical characteristics of chicory inulin and oligofructose

	Standard inulin	High performance inulin	Oligofructose 95
Chemical structure	GFn ($2 \leq n \leq 60$)	GFn ($10 \leq n \leq 60$)	GFn + Fn ($2 \leq n \leq 7$)
Average degree of polymerisation	12	25	4
Dry matter (%)	≥ 95	≥ 95	≥ 95
Inulin/oligofructose content (% on d.m)	92	≥ 99.5	95
Sugars content (% on d.m.)	8	≤ 0.5	5
pH (10% w/w)	5-7	5-7	5-7
Sulphated ash (% on d.m.)	< 0.2	< 0.2	< 0.2
Heavy metals (ppm on d.m.)	< 0.2	< 0.2	< 0.2
Appearance	White powder	White powder	White powder or colourless syrup
Taste	Neutral	Neutral	Moderately sweet
Sweetness (vs sucrose = 100%)	10%	None	35%
Solubility in water at 25°C (g/l)	120	< 25	> 750
Viscosity in water (5%) at 10°C (mPa.s)	1.6	2.4	< 1.0
Functionality in foods	Fat replacer	Fat replacer	Sugar replacer
Synergism	Synergy with gelling agents	Synergy with gelling agents	Synergy with intense sweeteners

G = glucosyl unit
F = fructosyl unit
d.m. = dry matter

1.1.2.2 Physical properties

Inulin has a bland neutral taste, without any off-flavour or aftertaste. Standard inulin is slightly sweet (10% sweetness in comparison with sugar), whereas high performance (long-chain) inulin is not sweet at all. It combines easily with other ingredients without modifying delicate flavours such as milk notes. It is moderately soluble in water (maximum 10% at room temperature) and has a rather low viscosity (less than 2 mPa.s for a 5% w/w solution in water) (2). On the other hand, inulin has a remarkable capacity to replace fat. When thoroughly mixed with water or another aqueous liquid, it forms a particle gel network resulting in a white creamy structure with a short spreadable texture that can easily be incorporated into foods to replace fat up to 100% (patented technology) (7). Such a gel is composed of a tri-dimensional network of insoluble sub-micron crystalline inulin particles in water. Large amounts of water are immobilised in that network, which assures its physical stability. As far as fat replacement is concerned, high performance inulin (patent-pending) shows about twice the functionality compared to standard chicory inulin. Special instant qualities that do not require shearing to give stable homogeneous gels also have been developed using a specific spray-drying process (patent-pending). Inulin works in synergy with some gelling agents, e.g. gelatine, alginate, k- and i-carrageenan, gellan gum and maltodextrins. It also improves the stability of foams and emulsions, such as aerated desserts, ice creams, table spreads and sauces (8).

Oligofructose is much more soluble than inulin (about 80% in water at room temperature). In its pure form it has a sweetness of about 35% in comparison with sucrose. Its sweetening profile closely approaches that of sugar. The taste is very clean without any lingering effect, and it even enhances fruit flavours. In combination with intense sweeteners such as sucralose, aspartame and acesulfame K, it provides interesting mixtures offering a rounder mouthfeel and a better sustained flavour, with reduced aftertaste. Combinations of acesulfame K-aspartame blends with oligofructose also exhibit a significant quantitative synergy (9). Oligofructose has good stability during normal food processing conditions (e.g. during heat treatment) even if the ß-bonds between the fructose units can be (partially) hydrolysed in very acid conditions. Fructose is formed in this process, which is more pronounced at low pH, high temperature and low dry substance conditions (2). Oligofructose also contributes to body and mouthfeel, shows humectant properties, reduces water activity ensuring high microbiological stability, and affects boiling and freezing points. So, in fact, it possesses technological properties that are closely related to those of sugar and glucose syrup (10).

1.1.3   Applications

Inulin and oligofructose can be used either for their nutritional advantages or their technological properties, but they are often applied to offer a double benefit: an improved organoleptic quality and a better-balanced nutritional composition. Table 1.1.III gives an overview of their applications in foods and drinks.

TABLE 1.1.III
Overview of food applications with inulin and oligofructose

Application	Functionality	Dosage level inulin (% w/w)	Dosage level oligofructose (% w/w)
Dairy products	Sugar & fat replacement Synergy with sweeteners Body & mouthfeel Fibre & prebiotic	2-5	2-5
Frozen desserts	Sugar & fat replacement Synergy with sweeteners Fibre & prebiotic	2-8	4-8
Table spreads (fat continuous)	Fat replacement Texture & spreadability Fibre & prebiotic	4-10	-
Baked goods & breads	Fibre & prebiotic Moisture retention Sugar replacement	2-15	2-25
Breakfast cereals	Fibre & prebiotic Crispness & expansion	2-25	2-15
Fillings	Sugar & fat replacement Texture improvement	2-30	2-50
Fruit preparations	Sugar replacement Synergy with sweeteners Body & mouthfeel Fibre & prebiotic	5-20	5-50
Salad dressings	Fat replacement Body & mouthfeel	2-10	-
Meat products	Fat replacement Texture & stability Fibre	2-10	-
Dietetic products & meal replacers	Sugar & fat replacement Synergy with sweeteners Low caloric value Body & mouthfeel Fibre & prebiotic	2-15	2-20
Chocolate	Sugar replacement Fibre	5-20	-
Tablets	Sugar replacement Fibre & prebiotic	5-100	2-10

The use of inulin or oligofructose as fibre ingredients is easy and often leads to an improved taste and texture (8). When used in bakery products and breakfast cereals, this presents major progress in comparison with classical dietary fibres.

8

Inulin and oligofructose give more crispness and expansion to extruded snacks and cereals, and they increase the bowl-life. They also keep breads and cakes moist and fresh for longer. Their solubility allows fibre incorporation in watery systems such as drinks, dairy products and table spreads. Inulin is also used as a dietary fibre in tablets. Inulin and oligofructose are also being used more commonly in functional foods, especially in a whole range of dairy products and baked goods such as breads, as prebiotic ingredients that selectively stimulate the growth of beneficial intestinal bacteria (11,12).

Thanks to its specific gelling characteristics, inulin enables the development of low-fat foods without compromising the taste or texture (7). This is particularly true in spreadable products such as table spreads, butter-like products, dairy spreads, cream cheeses, and processed cheeses. Inulin can replace significant amounts of fat and improve the stability of the emulsion, while providing a short spreadable texture. It gives excellent results in water-in-oil spreads with a fat content ranging from 20 to 60%, as well as in water-continuous formulations containing 15% fat or less (Figures 1.1.3 and 1.1.4). In low-fat dairy products such as milk drinks, fresh cheeses, yoghurts, creams, dips and dairy desserts, the addition of a few percent inulin imparts a better-balanced round flavour and a creamier mouthfeel. In dairy mousses (chocolate, fruit, yoghurt or fresh cheese-based), the incorporation of inulin improves the process ability and upgrades the quality. The resulting products retain their typical structure for a longer time (8). In frozen desserts, inulin facilitates easy processing, a rich creamy mouthfeel, excellent melting properties, as well as freeze-thaw stability. Fat replacement can be applied in meal replacers, meat products, sauces and soups. So, for instance, fat-reduced meat products with a creamier and juicier mouthfeel and an improved stability due to water immobilisation can be obtained. Inulin also has found an interesting application as a low calorie bulk ingredient in chocolate without added sugar, not to reduce fat, but to replace sugar, often in combination with a polyol such as isomalt.

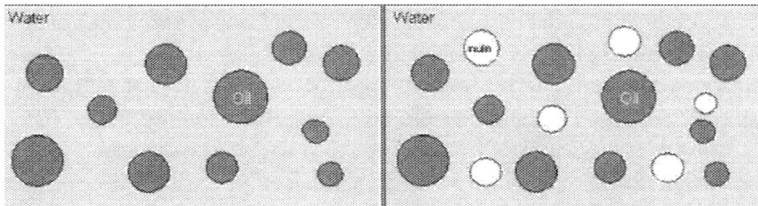

Fig.1.1.3. Inulin particles dispersed in an oil-in-water emulsion

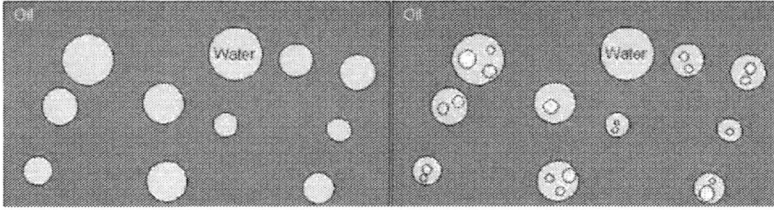

Fig.1.1.4. Inulin particles dispersed in a water-in-oil emulsion

In the yoghurt market, diet products are showing the strongest growth, in particular diet yoghurts with fruit. The incorporation of oligofructose (1 to 3%) in the recipe, often through the fruit preparation, improves the mouthfeel and offers a synergistic taste effect in combination with aspartame, acesulfame K and/or sucralose, without significantly increasing the caloric content. Oligofructose is also often formulated in other (low-calorie) dairy products, frozen desserts and meal replacers (13). Its incorporation in baked goods allows the replacement of sugar, fibre enrichment and better moisture retention properties. It also offers good binding characteristics in cereal bars. Its use is easy and requires only minor adaptation of the production process, if any. So, oligofructose is an ideal ingredient to give bulk with less calories and to provide nutritional benefits without compromising taste and mouthfeel.

Inulin and oligofructose are thus key ingredients offering new opportunities to the food industry looking for well-balanced and yet better-tasting products.

1.1.4 Physiological and nutritional properties

Thanks to their specific chemical structure (ß(2-1) bonds), which human digestive enzymes cannot hydrolyse, inulin and oligofructose pass from the mouth, to the stomach and the small intestine without undergoing any significant change, and without being metabolised. This has been confirmed in human studies with ileostomised volunteers (14). Inulin and oligofructose then enter the large intestine where they are totally fermented by the microbial flora, mainly by the beneficial bacteria. They are consequently transformed into bacterial mass, short-chain fatty acids (SCFA) and some gases (15). The fermentation of inulin is slower than that of oligofructose, allowing it to be active in more distal parts of the colon (16).

1.1.4.1 Dietary fibre effects

Due to their particular metabolic pathway, inulin and oligofructose are low calorie ingredients. Their caloric value is indeed limited to that contributed by the short chain fatty acids (SCFA) produced by fermentation, and partially metabolised in the body. Energy value studies have shown that inulin and oligofructose have a caloric value of around 1.5 kcal/g (or 6.3 kJ/g) (17). Thanks to their non-digestibility, inulin and oligofructose can furthermore be used by diabetics as their ingestion does not directly affect glycaemia, nor glucagon or insulin secretion.

Inulin and oligofructose on the other hand are soluble dietary fibres (18, 19). They are indeed plant components that are not hydrolysed by the human digestive enzymes and they induce typical fibre effects on the gut function, such as a reduction of stool pH, a relief of constipation and an increase in stool weight and frequency (or bulking effect). The mechanism behind these effects implies an increased bacterial biomass in the colon that results in an enhanced faecal bulk. The end result is a regularisation of bowel habits. The bulking index of inulin and oligofructose (g increase in stool weight per g fibre ingested) is of the order of 1.5-2 which is similar to other soluble fibres such as pectin or guar gum, as can be seen in Table 1.1.IV (20). In case of the ingestion of large amounts, typical side effects usually associated with dietary fibres (e.g. mild bloating) can also be observed.

TABLE 1.1.IV
Bulking effect of inulin, oligofructose and other fibres (20)

Fibre	Bulking index *
Oligofructose	1-2
Inulin	1.5-3.5
Pectin	1-2
Guar gum	1-2
Wheat bran	2.5-5

* g of increase in fresh faecal weight/g fibre eaten

1.1.4.2 Modulation of the gut microflora

The ingestion of inulin or oligofructose selectively stimulates the specific growth and/or metabolic activity of a limited number of beneficial bacteria in the colon, mainly bifidobacteria and lactobacilli, thereby contributing to the host's health. They are therefore "prebiotics" (21, 22).

The colon harbours a large population of microorganisms, composed of over 1,000 different species, which represent around 50% of the dry solids content of the large intestine. This microbial ecosystem plays a fundamental role in maintaining good health, and imbalances in the composition of the gut microflora have been linked to the aetiology of various diseases.

The prebiotic and bifidogenic effect of fructans has been demonstrated by using *in vitro* models (23-25), animals and, more importantly, human intervention studies (26-32). Human studies on the prebiotic effect of chicory inulin and oligofructose (from 5 to 20 g/day) have involved more than 400 volunteers (men and women, young adults and elderly) and used different study designs (1 to 12 weeks of supplementation, with a controlled diet or a free diet). They have demonstrated a major shift in intestinal bacterial composition, with bifidobacteria significantly increasing in numbers, whereas pathogens (e.g. clostridia) decreased in numbers, as illustrated in Figure 1.1.5. Molecular-based approaches, such as fluorescence *in situ* hybridisation (FISH), also confirmed that the specific quantitative increase in bifidobacteria counts as a consequence of inulin or oligofructose ingestion (30, 31).

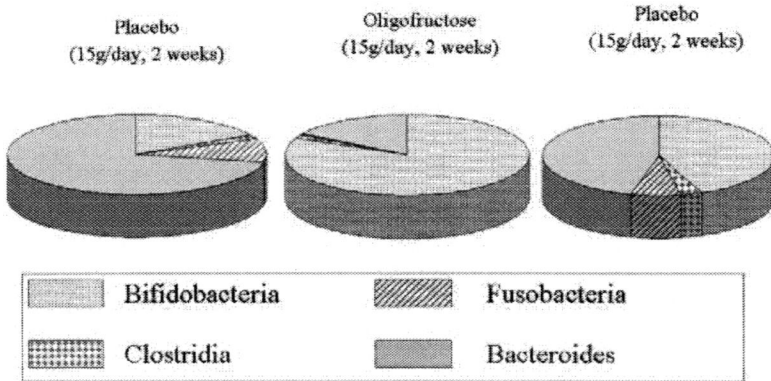

Placebo
(15g/day, 2 weeks)

Oligofructose
(15g/day, 2 weeks)

Placebo
(15g/day, 2 weeks)

	Bifidobacteria		Fusobacteria
	Clostridia		Bacteroides

Fig.1.1.5. Bifidus stimulation in humans (26)

Table 1.1.V overleaf summarises studies demonstrating the prebiotic effect of inulin and oligofructose in healthy adults.

Inulin and oligofructose both exert important bifidogenic properties and, based on available data, the quantitative increase in bifidobacteria counts seem to be rather independent of the fructan chain length. However, a difference in the profile and rate of fermentation appears between components with different chain length distribution. *In vitro* experiments have shown that the length of the fructan chain determines the rate of fermentation in the colon. Long-chain inulin fractions (DP > 10) are fermented about twice as slowly as low-DP molecules (DP < 10) (16). Furthermore, the impact of a combination of oligofructose (7.5 g/day) and long-chain inulin (7.5 g/day) on the mucosa-associated intestinal flora was studied in volunteers who underwent a colonoscopy. Biopsy samples were collected from the colonic mucosa, allowing quantification of the impact of the prebiotic on the mucosa-associated flora. The prebiotic intake significantly increased (mucosal) bifidobacteria and lactobacilli in both the proximal and distal parts of the colon (33).

TABLE 1.1.V
Prebiotic effect of inulin and oligofructose in healthy adult subjects

Reference	Dose Tested product	Number of volunteers (M/F)* / Age / Duration of supplementation	Increase in bifidobacteria (log)	Statistical significance
34	8 g/day Oligofructose	23 50-90 years 2 weeks	0.9	P<0.005
26	15 g/day Oligofructose	8 (7M/1F) 20-25 years 2 weeks	0.7	P<0.01
26	15 g/day Inulin	4 (4M) 20 years 2 weeks	0.9	P<0.001
35	12.5 g/day Oligofructose	20 (10M/10F) 22-39 years 12 days	1.2	P<0.01
36	4 g/day Oligofructose	12 (6M/6F) 20-34 years 25 days	1.0	P<0.03
27	20 and 40 g/day Inulin	10 (10F) 68-89 years 19 days	0.9 and 1.3	P<0.05
37	22-34 g/day Inulin	11 (6M/5F) 26-53 years 8.5 weeks	1.1	P<0.05
38	2.5-20 g/day Oligofructose	5x8 (18M/22F) 18-47 years 1 week	0.1-1.5 (according to dosage)	P<0.05 (starting from 5 g/day)
39	9 g/day Inulin	12 (12M) 22-24 years 4 weeks	0.3	P<0.05
28	8 g/day Oligofructose	8 (3M/5F) 20-50 years 2 et 5 weeks	0.8 and 1.0	P<0.01
29	5 g/day Oligofructose	8 (4M/4F) 20-40 years 11 days and 3 weeks	0.9 and 1.0	P<0.001
30	8 g/day Inulin	9 (M/F) 20-55 years 1 and 2 weeks	0.2	P<0.05
31	9 g/day Inulin	10 (6M/4F) 28-54 years 2 weeks	0.4	P=0.01
40	8 g/day Oligofructose	19 (4M/15F) 77-97 years 3 weeks	2.8	P<0.001

TABLE 1.1.V cont.
Prebiotic effect of inulin and oligofructose in healthy adult subjects

Reference	Dose Tested product	Number of volunteers (M/F)* / Age / Duration of supplementation	Increase in bifidobacteria (log)	Statistical significance
33	15 g/day Oligofructose and Inulin (1:1)	2x14 (8M/6F) 35-72 years 2 weeks	1.2	P=0.01
41	2.5-10 g/day Oligofructose	5x8 (M/F) 18-54 years 1 week	0.2-1.2 (according to dosage)	P<0.05
32	5 g/day Inulin	2x20 20-58 yrs. 2 and 4 weeks	1.0 and 1.3	P<0.001
42	7.7 and 15.4 g/day Inulin	3x15 (10M/35F) 21-26 years 1 and 2 weeks	0.5-1.2 (according to dosage)	P<0.05
43	8 g/day Oligofructose	12 (6M/6F) 67-71 years 4 weeks	0.7	P<0.05
44	5 and 8 g/day Inulin	30 (15M/15F) 19-35 years 2 weeks	0.2	P≤0.05
45	20 g/day Synergy1	19 (9M/10F) 21-23 years 4 weeks	0.8	P<0.001

* M: male
  F: female

Initial human studies have led to the development of "second generation" prebiotics. By combining a selected ratio of short-chain oligofructose and long-chain inulin, an oligofructose-enriched inulin with enhanced nutritional benefits was created (Orafti®Synergy1). The lower DP (oligofructose) fraction of Synergy1 is rapidly fermented, and hence this metabolic process is thought to take place in the proximal part of the colon. The long-chain fraction (inulin), on the other hand, is fermented at a slower rate and reaches more distal parts of the colon, which are primarily susceptible to colonic chronic disease. The end result is a wider spread and more sustained fermentation pattern with a prebiotic effect along the whole colon (46).

1.1.4.3 Enhanced resistance to infections and inflammation

The selective promoting effect of inulin and oligofructose on the bifidus population in the large intestine is usually regarded as beneficial to the well-being and health of the host (21). Its potential implications in the prevention and treatment of intestinal disorders and infections (e.g. bowel inflammation, irritable bowel syndrome, traveller's or antibiotic-associated diarrhoea) have been studied. By increasing the numbers of bifidobacteria, inulin and oligofructose impair the proliferation of potentially pathogenic bacteria, thus providing a better resistance towards enteric pathogens. This has been observed *in vitro*, as well as in animal and human studies.

It is clear from *in vitro* studies that lactic-acid producing bacteria (e.g. bifidobacteria and lactobacilli) can exert powerful antagonistic activities, thus inhibiting the growth of several pathogens (*Escherichia coli*, clostridia, *Campylobacter jejuni*, *Salmonella* or *Shigella*) (23, 47). These effects are mediated by the release of organic acids, which are end-products of inulin and oligofructose fermentation, as well as antimicrobial substances (bacteriocins). Organic acids (SCFA) lower the colonic pH that contributes to preventing overgrowth of pH-sensitive pathogenic bacteria. In addition, the fermentation of inulin and oligofructose exerts a trophic effect on the intestinal mucosa (48). Experiments with epithelial layers have also shown that these fructans are able to inhibit pathogen colonisation and support the intestinal barrier function (49).

In studies with gnotobiotic quails, inoculated with a flora from a baby with necrotising enterocolitis (comprising *Clostridium butyricum* and *Clostridium perfringens*) most of the quails died. On the contrary, additional colonisation of the quails with bifidobacteria led to their complete recovery and suppressed the growth of clostridia (50). Mice fed diets supplemented with inulin or oligofructose were better able to resist colonisation by *Candida albicans* than animals fed a control diet, the numbers of viable yeasts in their intestinal contents being lower in the supplemented groups (51). Inulin and oligofructose have also been shown to decrease systemic infections in rats, inducing a significantly lower mortality after an intra-peritoneal infection with virulent strains of *Listeria monocytogenes* or *Salmonella typhimurium*, as compared to control animals (51).

In humans, data from clinical trials in patients with gastrointestinal disorders have shown that inulin and oligofructose help to restore the balance of an altered gut microflora, and so accelerate the recovery of the gastrointestinal tract and also ameliorate disease symptoms.

In patients hospitalised for *Clostridium difficile*-associated diarrhoea (an infection triggered by antibiotherapy), supplementation with oligofructose increased bifidobacterial counts and decreased the relapses of diarrhoea, compared with the control group (52).

In young children (4-24 months of age) attending day-care centres, the supplementation with oligofructose (0.55 g per 15g of cereal) for 6 months resulted in significantly fewer episodes of fever and missed days of day-care because of diarrhoea, as well as a lower use of antibiotics for respiratory illnesses (53). These data were confirmed by another study conducted in young children receiving oligofructose (2 g/day), where greater numbers of faecal bifidobacteria were observed, concomitantly with lower numbers of clostridia. Also, episodes of fever, diarrhoea and vomiting were significantly reduced.

Improving the microbial balance of the intestinal ecosystem further showed promising implications in patients with inflammatory bowel disease, such as ulcerative colitis or Crohn's disease, in which the gut micro-ecology is affected. Studies in patients with ulcerative colitis have revealed that bifidobacterial populations in their colons are about 30-fold lower compared with those of healthy individuals (54). In patients with ulcerative colitis, supplementation with Synergy1 (and a probiotic) resulted in decreased inflammation, as well as partially regenerated epithelial tissue (55). These benefits were accompanied by higher mucosal levels of bifidobacteria, resulting in a 42-fold increase compared with the control group. Decreased disease activity and improved recovery and/or remission were also observed following the intake of inulin and oligofructose in patients with Crohn's disease (56). Reduction of inflammation and associated factors after inulin consumption was further shown in patients with pouchitis (57).

1.1.4.4 Improved calcium absorption

In rats, the ingestion of inulin or oligofructose has repeatedly been shown to significantly increase the absorption of important minerals such as calcium and magnesium, and even to improve bone mineralisation (bone mineral density) and to prevent ovariectomy-induced bone resorption (58-60). These findings suggest that inulin-type fructans have a beneficial effect on bone loss caused by oestrogen deficiency. Oligofructose was also shown to improve the biomechanical properties of bones (femur resistance to fracture) in healthy growing rats (61). A direct comparison of several fructans of different chain lengths has indicated that the highest increase in calcium absorption occurs with oligofructose-enriched inulin (Synergy1) combining both short and longer fructan chains, compared to inulin or oligofructose given alone (62).

These data highlight the important impact that inulin and oligofructose (mainly Synergy1) can have for the prevention of osteoporosis. Increased absorption of calcium has been confirmed in humans (adults, adolescents and post-menopausal women) receiving chicory inulin and oligofructose in their diet, more specifically oligofructose-enriched inulin (Synergy1), as shown in Table 1.1.VI. In an early study with healthy adult volunteers who were given 40 g inulin/day, a significant increase in calcium absorption was observed using the balance technique (63). A subsequent study carried out in adolescent boys with oligofructose supplementation (15 g/day) found a significant increase (26%) in true calcium absorption. Calcium absorption was measured through the method of dual stable isotopes (^{46}Ca/^{42}Ca) (64). Further studies were conducted in adolescent girls with adequate calcium intakes, demonstrating once again a significant increase in true calcium absorption, with as low as 8 g/day consumption of Synergy1 compared with the placebo group (65, 66). More in-depth evaluation of the subjects revealed that the adolescents who had a lower level of calcium absorption while on placebo had the greatest increase in calcium absorption in response to Synergy1, and thus seemed to benefit the most from the prebiotic supplementation. It is also worth noting that at such a low dose (8 g/day) of oligofructose did not significantly influence the level of calcium absorption (65). Lastly, positive effects were also observed in post-menopausal women, supplemented with 10 g/day of either

Synergy1 or a placebo, resulting in a significant increase in both calcium and magnesium absorption. The treatment was also found to positively affect markers of bone turnover. The efficacy of the intervention seemed to be even more important in women with lower initial bone mineral density of the lumbar spine (67).

TABLE 1.1.VI
Effects of inulin and oligofructose on calcium absorption in healthy individuals

Reference	Dose Tested product	Number of volunteers / (M/F)* Age / Duration of supplementation	Calcium absorption with control (%)	Calcium absorption with test (%) (+ relative increase)	Statistical significance
63	40 g/day Inulin	9 (9M) 19-24 years 26 days	21.3	33.7 (+58%)	P<0.01
68	15 g/day Oligofructose or Inulin	12 (12M) 20-30 years 3 weeks	28.1 (only 24h urine collection)	26.3 (Oligofructose) 25.8 (Inulin)	NS
64	15 g/day Oligofructose	12 (12M) 14-16 years 9 days	47.8 (36h urine collection)	60.1 (+26%)	P<0.05
65	8 g/day Oligofructose	30 (30F) 11-14 years 3 weeks	31.8	31.8	NS
65	8 g/day Synergy1	29 (29F) 11-14 years 3 weeks	32.3	38.2 (+18%)	P<0.01
66	8 g/day Synergy1	54 (54F) 10-15 years 3 weeks	33.1	36.1 (+9%)	P<0.05
69	10 g/day Oligofructose	12 (12F) 50-70 years 5 weeks	35.6	36.6	NS
70	8 g/day Synergy1	100 (50M/50F) 9-13 years 8 weeks and 1 year	30.0 (at 8-week) 31.7 (at 1-year)	38.5 (at 8 week) (+28%) 37.7 (at 1 year) (+19%)	P<0.05
67	10 g/day Synergy1	15 (15F) 66-79 years 6 weeks	22.2	27.3 (+23%)	P<0.05

* M: male
  F: female

A beneficial impact on bone mineralisation was demonstrated in a long-term study conducted in pubertal girls and boys (Tanner stage 2 and 3). Calcium accretion is at its optimal rate during puberty and its level may impact bone health at a later age. A total of 100 adolescents (from 9 to 13 years old) were involved in this 1-year randomised, double-blinded intervention study, and were supplemented with 8 g/day of either Synergy1 or placebo (maltodextrin). True calcium absorption was significantly enhanced after 8 weeks in the Synergy1 group and this beneficial effect was maintained during the whole intervention year (Figure 1.1.6). At the end of the experimental period, supplementation with Synergy1 resulted in a higher change in whole-body bone mineral density and whole-body bone mineral content. Under the assumption that the fraction of calcium in bone mineral is about 32%, these values correspond to an additional net accretion of 30 mg of calcium per day, with Synergy1 supplementation (70).

A recent kinetic study has further confirmed that such increased calcium absorption in response to Synergy1 intake primarily originates in the colon (71).

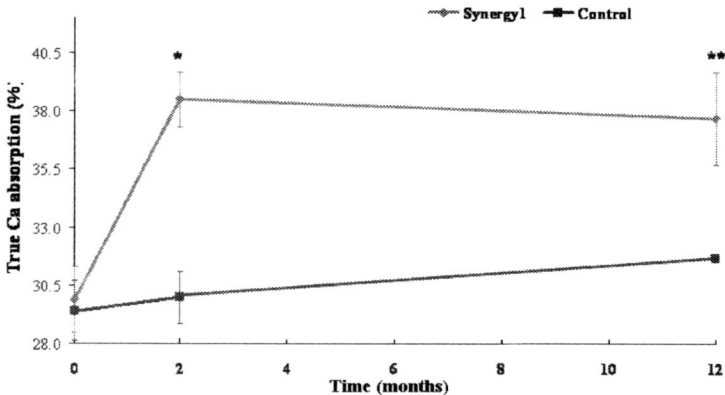

Significantly different from placebo   *(p<0.001)   **(p<0.05)

Fig. 1.1.6. Effect of Synergy1 (8g/day) on calcium absorption in adolescents (70)

1.1.4.5   *Modulation of appetite, food intake and body weight gain*

Appetite is regulated by complex mechanisms, including hormones released by the gastrointestinal tract and peripheral tissues in response to components of our diet. These hormones inform the brain (primarily the hypothalamus) of our feeding state, and these signals are translated into feelings of hunger and/or being full.

Appetite-suppressing peptides include cholecystokinin (CCK), PP-fold peptide (PYY) and glucagon-like peptide-1 (GLP-1). Higher circulating levels of these hormones are associated with lower subjective hunger ratings and lower food intake. In contrast, blood levels of the hormone ghrelin are increased in the fasting state, inducing hunger and subsequently initiating food intake.

Experimental data have recently accumulated, demonstrating that dietary fructans, and in particular oligofructose and Synergy1, are able to modulate the secretion and subsequent release into the blood circulation of gut hormones that are involved in appetite regulation. Amongst these hormones, glucagon-like peptide-1 (GLP-1) and grehlin have been investigated in animal studies. GLP-1 is a peptide released from the entero-endocrine L-cells present in the ileum and the colon, in response to nutrient ingestion. In various animal models e.g. rats on a normal or high-fat diet, obese and diabetic rats, it was demonstrated that oligofructose or Synergy1 significantly enhanced the GLP-1 content in the colon and/or in portal blood, resulting in lower food and energy intake, as well as lower body weight gain and adipose tissue development (72-74). The plasma levels of grehlin were lower in oligofructose and Synergy1-fed rats. Observed effects on body weight are persistent, as shown by a life-long intervention study in rats where Synergy1 supplementation resulted in a lower body weight evolution during the whole life-span of the animals (75).

The mechanism by which oligofructose and Synergy1 increase the production of GLP-1 was hypothesised to involve SCFA produced in the colon as a consequence of fructan fermentation. Evidence is emerging about their role in human physiology, e.g. lipid metabolism. The production of GLP-1 might obviously constitute a link between the outcome of fermentation in the colon and the modulation of food intake. The role of GLP-1 was tested in mice using a GLP-1 receptor antagonist. Mice fed a high-fat diet were supplemented with oligofructose and exhibited lower food intake and body weight gain, as well as anti-diabetic effects. However the treatment of the mice with the GLP-1 receptor antagonist, on the contrary, totally prevented the beneficial effects seen with oligofructose. The importance of GLP-1 in mediating the effects of oligofructose was also confirmed through the use of GLP-1 receptor knock-out mice (GLP-1R -/-) that were totally insensitive to the effects of oligofructose on body weight, food intake and parameters of glucose metabolism.

The effects of inulin-type fructans on appetite regulation and energy intake were ultimately investigated in humans, in a placebo-controlled, single-blinded intervention study involving 10 healthy volunteers receiving oligofructose or a placebo (8 g twice daily). Oligofructose intake resulted in increased satiety, reduced hunger and prospective food consumption (assessed by visual analogue scales). This led to a lower total energy intake during the day, as illustrated in Figure 1.1.7 (76). Data from a 1-year intervention trial in adolescents (n=100) further showed that the administration of Synergy1 (8 g/day) resulted in a significantly lower body mass index (BMI), lower body weight gain and lower body fat mass, thus supporting adequate weight maintenance during early adolescence (77).

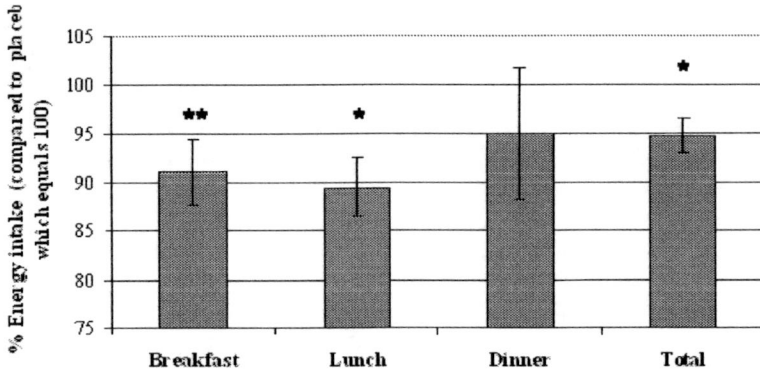

Significantly different from placebo   *(p≤0.05)   **(p ≤ 0.01)

Fig. 1.1.7. Effect of oligofructose (16g/day) on energy intake in healthy adults (76)

1.1.4.6   Lipid metabolism

Animal studies have repeatedly shown a very significant reduction (-40%) in serum and liver triglycerides (and possibly an improvement in the HDL/LDL cholesterol ratio) upon ingestion of inulin or oligofructose (78-81). *In vitro* studies with isolated hepatocytes have shown that inulin and oligofructose decrease the activity of key enzymes involved in lipogenesis, thus lowering the *de novo* synthesis of fatty acids in the liver (82). A corresponding reduction in the gene expression of these enzymes was shown. The hypo-triglyceridemic effect of inulin and oligofructose (or their fermentation metabolites) can also be explained by their impact on several hormones (insulin and/or glucose-dependent insulinotropic polypeptide or glucagon-like peptide-1) (83). An improvement in the lipid metabolism has also been observed in different human studies with inulin, in particular in (slightly) hyper-lipidemic subjects (84-86). A recent meta-analysis has confirmed the significant serum triglyceride-lowering effect of inulin and oligofructose (87).

The influence of inulin and oligofructose was also studied in models of unbalanced diets, a common problem in our Western society. Oligofructose (10%) supplementation in rats consuming a high-fat diet resulted in decreased post-prandial serum triglycerides and cholesterol (88). In the same experimental model, in addition to the lowering effect on serum triglycerides, oligofructose decreased body weight gain and adipose tissue development (74). Further studies were conducted with genetically obese (fa/fa Zucker) rats, fed a diet supplemented with 10% of oligofructose or oligofructose-enriched inulin (Synergy1). A protective effect towards hepatic steatosis was demonstrated, the hepatic triglyceride concentration being reduced by more than 50% compared with the control rats. In addition, body weight and body fat mass were significantly reduced, an effect that was not observed when non-fermentable fibres (e.g. cellulose) were added to the diet (89, 72). Elevated serum triglycerides and cholesterol levels have been highlighted as risk factors for coronary heart disease. In a model of atherosclerosis (in apoE-deficient mice), inulin and oligofructose (10%) significantly inhibited atherosclerotic plaque formation (in the aorta), concomitantly with a reducing effect on cholesterol and triglycerides levels (90).

1.1.4.7 Reduction of colon cancer risk

In experimental animals, inulin and oligofructose decrease the incidence of pre-cancerous lesions (aberrant crypt foci) as well as cancerous tumours in the colon (91, 92). They could therefore contribute to the reduction of risk for colon cancer in humans. It is also worth noting that, as shown in Figure 1.1.8, a symbiotic synergistic protective effect has been found for the combination of high performance inulin and bifidobacteria (93). The beneficial effects for colon cancer prevention appear to be more pronounced with long-chain inulin-type fructans, such as high performance inulin and Synergy1 (94). Indeed, cancerous lesions are more predominant in the distal regions of the colon, where the long-chain fractions are fermented and exert their prebiotic properties.

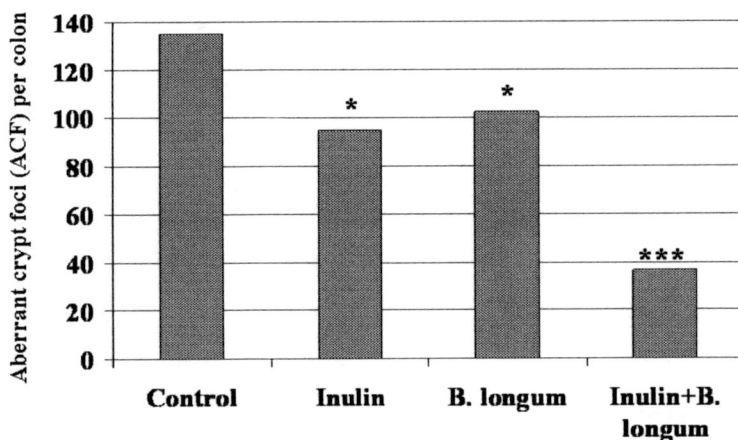

* Significantly different from control (p<0.05)
*** Significantly different from control (p<0.001)

Fig. 1.1.8. Effect of inulin and bifidobacteria on colonic pre-cancerous lesions (ACF) in rats (93)

Lately, a phase-II anti-cancer study was conducted involving 80 volunteers with a history of colorectal cancer or at increased risk of colon cancer (polypectomised volunteers). Synergy1, in combination with probiotics, was administered for 12 weeks in a randomised, double-blind, placebo-controlled design. The synbiotic resulted in a significant reduction in DNA damage and colonic cell proliferation rate. These benefits coincided with significantly higher levels of bifidobacteria and lactobacilli, as well as a decrease in the numbers of pathogens (coliforms and *Clostridium perfringens*). It was hypothesised that these changes in bacterial profiles beneficially altered the metabolic activity in the colon. Also, a decrease in inflammation and exposure of the epithelium to damaging agents was observed. This is important since mucosal inflammation, as well as high cytotoxicity and genotoxicity of the luminal contents are associated with an increased risk of colon cancer. Altogether these effects provide evidence of a possible protective effect of fructan-based synbiotics against colon cancer development (95).

As more and more scientific data become available, the nutritional and health benefits of the prebiotics inulin and oligofructose, as summarised in Table 1.1.VII, continue to surprise researchers and nutritionists. Chicory inulin and oligofructose are therefore often taken as a practical illustration of active food ingredients designated 'functional foods'.

TABLE 1.1.VII
Well-established nutritional properties of inulin and oligofructose

Non-digestibility and low caloric value (1.5 kcal/g)
Suitable for diabetics
Soluble dietary fibre
Stool bulking effect : increase in stool weight and stool frequency
Modulation of the gut flora, promoting beneficial bacteria (bifidobacteria) and repressing harmful ones (clostridia) : prebiotic effect
Improvement of calcium bioavailability and bone mineralisation

Emerging nutritional properties of inulin and oligofructose

Protection against intestinal disorders, infections and inflammation
Modulation of appetite and energy intake
Reduction in serum triglycerides
Reduction of biomarkers of colon cancer risk

1.1.5   Analytical methods

At the moment, the authorities in most countries have confirmed that inulin and oligofructose can be labelled as 'dietary fibre' for food labelling. However, the classical methods for analysis of dietary fibres do not analyse them. Based on the results of a collaborative ring test, the AOAC International has adopted as method number 997.08 the "Fructan method" that specifically allows the accurate quantitative determination of inulin and oligofructose in foods (96). The method, represented in Figure 1.1.9, involves the treatment of the sample with amylo-glucosidase and inulinase enzymes, followed by the determination of the released sugars by ion exchange chromatography. This method can be combined with the standard AOAC 'Total Dietary Fibre' method to produce the total amount of fibre to be used for food labelling.

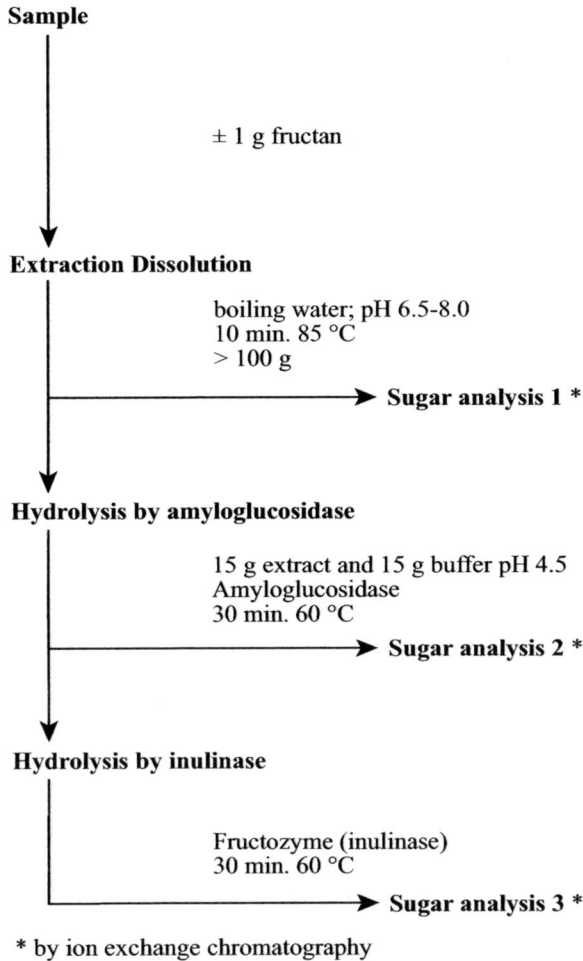

Sample

$\pm$ 1 g fructan

Extraction Dissolution

boiling water; pH 6.5-8.0
10 min. 85 °C
> 100 g

→ **Sugar analysis 1 ***

Hydrolysis by amyloglucosidase

15 g extract and 15 g buffer pH 4.5
Amyloglucosidase
30 min. 60 °C

→ **Sugar analysis 2 ***

Hydrolysis by inulinase

Fructozyme (inulinase)
30 min. 60 °C

→ **Sugar analysis 3 ***

* by ion exchange chromatography

Adapted from Hoebregs H. (96)

Fig.1.1.9. Analytical 'Fructan method'

1.1.6   References

1. Van Loo J., Coussement P., De Leenheer L., Hoebregs H., Smits G. On the presence of inulin and oligofructose as natural ingredients in the Western diet. *Critical Reviews in Food Science and Nutrition*, 1995, 35 (6), 525-52.

2. De Leenheer L. Production and use of inulin: Industrial reality with a promising future, in *Carbohydrates as organic raw materials III*, VCH Publ. Inc., New York, Eds Van Bekkum H., Röper H., Voragen A.G.J., 1996, 67-92.

3. Schöne A. Betrachtungen über Verwendungsmöglichkeit der Zichorie. *Centralblat für die Zuckerindustrie,*. 1920, 27, 396-8.

4. Belval H. Industrie de l'inuline et du lévulose, in *Dix ans d'efforts scientifiques, industriels et coloniaux 1914-1924*, Chimie et Industrie, Paris, 1927, 1068-9.

5. Bornet F.R.J. Undigestible sugars in food products. *American Journal of Clinical Nutrition*, 1994, 59, 763S-9S.

6. De Leenheer L., Hoebregs H. Progress in the elucidation of the composition of chicory inulin. *Starch,*. 1994, 46 (5), 193-6.

7. Franck A. Rafticreming: The new process allowing to turn fat into dietary fiber, in *FIE Conference proceedings*, Expoconsult Publishers, Maarssen, 1993, 193-7.

8. Franck A., Coussement P. Multi-functional inulin. *Food ingredients and Analysis International*, 1997, 8-10.

9. Wiedmann M., Jager M. Synergistic sweeteners. *Food Ingredients and Analysis International*, 1997, 51-6.

10 Crittenden R.G., Playne M.J. Production, properties and applications of food-grade oligosaccharides. *Trends in Food Science and Technology,*. 1996, 7, 353-61.

11. Coussement P. Pre-and synbiotics with inulin and oligofructose. *Food Technology Europe*. 1996, 102-4.

12. Walter T. Bread goes prebiotic. *International Food Ingredients,*. 1999, 2, 20-1.

13. Franck A. Prebiotic sweeteners blends. *Food Marketing and Technology*. 1999, 13 (1), 22-4.

14. Ellegard L., Andersson H., Bosaeus I. Inulin and oligofructose do not influence the absorption of cholesterol, or the excretion of cholesterol, Ca, Mg, Zn, Fe, or bile acids but increase energy excretion in ileostomy subjects. *European Journal of Clinical Nutrition*, 1997, 51, 1-5.

15. Delzenne N.M., Roberfroid M.R. Physiological effects of non-digestible oligosaccharides. *Lebensmittel.-Wissenschaft und-Technologie*. 1994, 27, 1-6.

16. Roberfroid M.B., Van Loo J., Gibson G.R. The bifidogenic nature of chicory inulin and its hydrolysis products. *The Journal of Nutrition,*. 1998, 128, 11-19.

17. Roberfroid M., Gibson G.R., Delzenne N. The biochemistry of oligofructose, a nondigestible fiber: An approach to calculate its caloric value. *Nutrition Reviews,*. 1993, 51 (5), 137-46.

18. Roberfroid M. Dietary fiber, inulin and oligofructose: A review comparing their physiological effects. *Critical Reviews in Food Science and Nutrition*, 1993, 33 (2), 103-48.

19. Prosky L. Inulin and oligofructose are part of the dietary fiber complex. *Journal of AOAC International,*. 1999, 82 (2), 223-6.

20. Roberfroid M.B. Health benefits of non-digestible oligosaccharides, in *Dietary Fiber in Health and Disease*, Eds. Kritchevsky D., Bonfield C., New York, Plenum Press, 1997, 211-9.

21. Gibson G.R., Roberfroid M.B. Dietary modulation of the human colonic microbiota – Introducing the concept of prebiotics. *The Journal of Nutrition*, 1995, 125, 1401-12.

22. Gibson G.R., Probert H.M., Van Loo J., Rastall R.A., Roberfroid M.B. Dietary modulation of the human colonic microbiota: updating the concept of prebiotics. *Nutrition Research Reviews*, 2004, 17, 259-75.

23. Wang X., Gibson G.R. Effects of the *in vitro* fermentation of oligofructose and inulin by bacteria growing in the human large intestine. *Journal of Applied Bacteriology*, 1993, 75, 373-80.

24. Gibson G.R., Wang X. Bifidogenic properties of different types of fructo-oligosaccharides. *Food Microbiology*, 1994, 11, 491-8.

25. Hopkins M.J., Cummings J.H., Macfarlane G.T. Inter-species differences in maximum specific growth rates and cell yields of bifidobacteria cultured on oligosaccharides and other simple carbohydrate sources. *Journal of Applied Microbiology*, 1998, 85, 381-6.

26. Gibson G.R., Beatty E.R., Wang X., Cummings J.H. Selective stimulation of bifidobacteria in the human colon by oligofructose and inulin. *Gastroenterology*, 1995, 108, 975-82.

27. Kleessen B., Sykura B., Zunft H.J., Blaut M. Effects of inulin and lactose on fecal microflora, microbial activity, and bowel habit in elderly constipated persons. *American Journal of Clinical Nutrition*, 1997, 65, 1397-1402.

28. Menne E., Guggenbuhl N., Roberfroid M. Fn-type chicory inulin hydrolysate has a prebiotic effect in humans. *The Journal of Nutrition*, 2000, 130, 1197-9.

29. Rao V.A. The prebiotic properties of oligofructose at low intake levels. *Nutrition Research,*. 2001, 21, 843-8.

30. Tuohy K.M., Finlay R.K., Wynne A.G., Gibson G.R. A human volunteer study on the prebiotic effects of HP-inulin - faecal bacteria enumerated using Fluorescent *In Situ* Hybridisation (FISH). *Anaerobe*, 2001, 7, 113-8.

31. Harmsen H.J.M., Raangs G.C., Franks A.H., Wildeboer-Veloo A.C.M., Welling G.W. The effect of the prebiotic inulin and the probiotic *Bifidobacterium longum* on the fecal microflora of healthy volunteers measured by FISH and DGGE. *Microbial Ecology in Health and Disease*, 2002, 14, 211 - 9.

32. Bouhnik Y., Raskine L., Champion K., Andrieux C., Penven S., Jacobs H., Simoneau G. Prolonged administration of low-dose inulin stimulates the growth of bifidobacteria in humans. *Nutrition Research,*. 2007, 27, 187-93.

33. Langlands S.J., Hopkins M.J., Coleman N., Cummings J.H. Prebiotic carbohydrates modify the mucosa associated microflora of the human large bowel. *Gut*, 2004, 53, 1610-6.

34. Mitsuoka T., Hidaka Y., Eida T. Effect of fructooligosaccharides on intestinal microflora. *Die Nahrung*, 1987, 31, 5-6.

35. Bouhnik Y., Flourié B., Riottot M., Bisetti N., Gailing M-F., Guibert A., Bornet F., Rambaud J-C. Effects of fructo-oligosaccharides ingestion on fecal bifidobacteria and selected metabolic indexes of colon carcinogenesis in healthy humans. *Nutrition and Cancer*, 1996, 26, 21-29.

36. Buddington R.K., Williams C.H., Chen S-C., Witherly S.A. Dietary supplement of neosugar alters the fecal flora and decreases activities of some reductive enzymes in human subjects. *American Journal of Clinical Nutrition*, 1996, 63, 709-16.

37. Kruse H-P., Kleessen B., Blaut M. Effects of inulin on faecal bifidobacteria in human subjects. *British Journal of Nutrition*, 1999, 82, 375-82.

38. Bouhnik Y., Vahedi K., Achour L., Attar A., Salfati J., Pochart P., Marteau P., Flourié B., Bornet F., Rambaud J-C. Short-chain fructo-oligosaccharide administration dose-dependently increases fecal bifidobacteria in healthy humans. *Journal of Nutrition*, 1999, 129, 113-16.

39. Brighenti F., Casiraghi M-C., Canzi E., Ferrari A. Effect of consumption of a ready-to-eat breakfast cereal containing inulin on the intestinal milieu and blood lipids in healthy volunteers. *European Journal of Clinical Nutrition*, 1999, 53, 726-33.

40. Guigoz Y., Rochat F., Perruisseau-Carrier G., Rochat I., Schriffin E.J. Effects of oligosaccharide on the faecal flora and non-specific immune system in elderly people. *Nutrition Research*, 2002, 22, 13-25.

41. Bouhnik Y., Raskine L., Simoneau G., Vicaut E., Neut C., Flourié B., Brouns F., Bornet F. The capacity of nondigestible carbohydrates to stimulate faecal bifidobacteria in healthy humans: a double-blind, randomized, placebo-controlled, parallel-group, dose-response relation study. *American Journal of Clinical Nutrition*, 2004, 80, 1658-64.

42. Kleessen B., Schwarz S., Boehm A., Fuhrmann H., Richter A., Henle T., Krueger M. Jerusalem artichoke and chicory inulin in bakery products affect faecal microbiota of healthy volunteers. *British Journal of Nutrition*, 2007, 98, 540-49.

43. Bouhnik Y., Achour L., Paineau D., Riottot M., Attar A., Bornet F. Four-week short chain fructo-oligosaccharides ingestion leads to increasing fecal bifidobacteria and cholesterol excretion in healthy elderly volunteers. *Nutrition Journal*, 2007, Dec. 5, 6, 42.

44. Kolida S., Meyer D., Gibson G.R. A double-blind placebo-controlled study to establish the bifidogenic dose of inulin in healthy humans. *European Journal of Clinical Nutrition*, 2007, 61(10), 189-95.

45. De Preter V., Vanhoutte T., Huys G., Swings J., Rutgeerts P., Verbeke K. Baseline microbiota activity and initial bifidobacteria counts influence responses to prebiotic dosing in healthy subjects. *Alimentary pharmacology & therapeutics*, 2008, 27(6), 504-13.

46. Van Loo J. The specificity of the interaction with intestinal bacterial fermentation by prebiotics determines their physiological efficacy. *Nutrition Research Reviews*,. 2004, 17, 89-98.

47. Fooks L.J., Gibson G.R. *In vitro* investigations of the effect of probiotics and prebiotics on selected human intestinal pathogens. *FEMS Microbiology Ecology*,. 2002, 39, 67-75.

48. Kleessen B., Hartmann L., Blaut M. Fructans in the diet cause alterations of intestinal mucosal architecture, released mucins and mucosa-associated bifidobacteria in gnotobiotic rats. *British Journal of Nutrition*, 2003, 89, 597-606.

49. Naughton P.J., Mikkelsen L.L., Jensen B.B. Effects of nondigestible oligosaccharides on *Salmonella enterica* serovar typhimurium and non-pathogenic *Escherichia coli* in the pig small intestine *in vitro. Applied and Environmental Microbiology*, 2001, 67 (8), 3391-5.

50. Butel M.J., Roland N., Hibert A., Popot F., Favre A., Tessedre A.C., Bensaada M., Rimbault A., Szylit O. Clostridial pathogenicity in experimental necrotising enterocolitis in gnotobiotic quails and protective role of bifidobacteria. *Journal of Medical Microbiology*, 1998, 47, 391-9.

51. Buddington K.K., Donahoo J.B., Buddington R.K. Dietary oligofructose and inulin protect mice from enteric and systemic pathogens and tumor inducers. *The Journal of Nutrition*, 2002, 132, 472-7.

52. Lewis S., Burmeister S., Brazier J. Effect of the prebiotic oligofructose on relapse of *Clostridium difficile*-associated diarrhea: a randomized, controlled study. *Clinical Gastroenterology and Hepatology*, 2005, 3, 442-8.

53. Saavedra J.M., Tschernia A. Human studies with probiotics and prebiotics: clinical implications. British Journal of Nutrition. 2002, 87 (Suppl. 2), S241-6.

54. Macfarlane S., Furrie E., Kennedy A., Cummings J.H., Macfarlane G.T. Mucosal bacteria in ulcerative colitis. *British Journal of Nutrition*. 2005, 93, (Suppl.1), S67-72.

55. Furrie E., Macfarlane S., Kennedy A., Cummings J.H., Walsh S.V., O'Neil D.A., Macfarlane G.T. Synbiotic therapy (*Bifidobacterium longum*/ Synergy1) initiates resolution of inflammation in patients with active ulcerative colitis: a randomised controlled pilot trial. *Gut*, 2005, 54, 242-9.

56. Lindsay J.O., Whelan K., Stagg A.J., Gobin P., Al-Hassi H.O., Rayment N., Kamm M.A., Knight S.C., Forbes A. Clinical, microbiological, and immunological effects of fructo-oligosaccharide in patients with Crohn's disease. *Gut*, 2006, 55, 348-55.

57. Welters C.F.M., Heineman E., Thunnissen F.B.J.M., Van den Bogaard A.E.J.M., Soeters P.B., Baeten C.G.M.I. Effect of dietary inulin supplementation on inflammation of pouch mucosa in patients with an ileal pouch-anal anastomosis. *Rectum*, 2002, 45, 621-7.

58. Delzenne N., Aertssens J., Verplaetse N., Roccaro M. & Roberfroid M. Effect of fermentable fructo-oligosaccharides on energy and nutrients absorption in the rat. *Life Science*, 1995, 57 (17), 1579-87.

59. Scholz-Ahrens K.E., Açil Y., Schrezenmeir J. Effect of oligofructose or dietary calcium on repeated calcium and phosphorus balances, bone mineralization and trabecular structure in ovariectomized rats. *British Journal of Nutrition*, 2002, 88, 365-77.

60. Zafar T.A., Weaver C.M., Zhao Y., Martin B.R., Wastney M.E. Nondigestible oligosaccharides increase calcium absorption and suppress bone resorption in ovariectomized rats. *The Journal of Nutrition*, 2004, 134, 399-402.

61. Lobo A.R., Colli C., Filisetti T.M.C.C. Fructooligosaccharides improve bone mass and biomechanical properties in rats. *Nutrition Research*, 2006, 26, 413-20.

62. Coudray C., Tressol J.C., Gueux E., Rayssiguier Y. Effects of inulin-type fructans of different chain length and type of branching on intestinal absorption and balance of calcium and magnesium in rats. *European Journal of Nutrition*, 2003, 42, 91-8.

63. Coudray C., Bellanger J., Castiglia-Delavaud C., Rémésy C., Vermorel M., Rayssiguier Y. Effect of soluble or partly soluble dietary fibre supplementation on absorption and balance of calcium, magnesium, iron and zinc in healthy young men. *European Journal of Clinical Nutrition*, 1997, 51, 375-80.

64. Van den Heuvel E., Muys T., van Dokkum W., Schaafsma G. Oligofructose stimulates calcium absorption in adolescents. *American Journal of Clinical Nutrition*, 1999, 69, 544-8.

65. Griffin I.J., Davila P.M., Abrams S.A. Non-digestible oligosaccharides and calcium absorption in girls with adequate calcium intakes. *British Journal of Nutrition*. 2002, 87 (Suppl. 2), S187-S91.

66. Griffin I.J., Hicks P.M.D., Heaney R.P., Abrams S.A. Enriched chicory inulin increases calcium absorption mainly in girls with lower calcium absorption. *Nutrition Research*, 2003, 23, 901-9.

67. Holloway L., Moynihan S., Abrams S.A., Kent K., Hsu A.R., Friedlander A.L. Effects of oligofructose-enriched inulin on intestinal absorption of calcium and magnesium and bone turnover markers in postmenopausal women. *British Journal of Nutrition*, 2007, 97, 365-72.

68. Van den Heuvel E., Schaafsma G., Muys T., van Dokkum W. Nondigestible oligosaccharides do not interfere with calcium and nonheme-iron absorption in young, healthy men. *American Journal of Clinical Nutrition*, 1998, 67, 445-51.

69. Tahiri M., Tressol J.C., Arnaud J., Bornet F.R., Bouteloup-Demange C., Feillet-Coudray C., Brandolini M., Ducros V., Pépin D., Brouns F., Roussel A.M., Rayssiguier Y., Coudray C. Effect of short-chain frutooligosaccharides on intestinal calcium absorption and calcium status in postmenopausal women: a stable-isotope study. *American Journal of Clinical Nutrition*, 2003, 77, 449-57.

70. Abrams S.A., Griffin I.J., Hawthorne K.M., Liang L., Gunn S.K., Darlington G., Ellis K.J. A combination of prebiotic short- and long-chain inulin-type fructans enhances calcium absorption and bone mineralization in young adolescents. *American Journal of Clinical Nutrition*, 2005, 82, 471- 6.

71. Abrams S.A., Hawthorne K.M., Aliu O., Hicks P.D., Chen Z., Griffin I.J. An inulin-type fructan enhances calcium absorption primarily via an effect on colonic absorption in humans. *The Journal of Nutrition*, 2007, 137, 2208-12.

72. Daubioul C., Rousseau N., Demeure R., Gallez B., Taper H., Declerck B., Delzenne N. Dietary fructans, but not cellulose, decrease triglyceride accumulation in the liver of obese Zucker *fa/fa* rats. *The Journal of Nutrition*, 2002, 132, 967-73.

73. Cani P.D., Dewever C., Delzenne N.M. Inulin-type fructans modulate gastrointestinal peptides involved in appetite regulation (glucagon-like peptide-1 and ghrelin) in rats. *British Journal of Nutrition*, 2004, 92, 521-6.

74. Cani P.D., Neyrinck A.M., Maton N., Delzenne N. Oligofructose promotes satiety in rats fed a high-fat diet: involvement of glucagon-like peptide-1. *Obesity Research*, 2005, 13 (6), 1000-7.

75. Rozan P., Nejdi A., Hidalgo S., Bisson J-F., Desor D., Messaoudi M. Effects of lifelong intervention with an oligofructose-enriched inulin in rats on general health and lifespan. *British Journal of Nutrition*, 2008 (in press), published online ahead of print, 11 Apr., doi: 10.1017/S0007114508975607.

76. Cani P.D., Joly E., Horsmans Y., Delzenne N.M. Oligofructose promotes satiety in healthy human: a pilot study. *European Journal of Clinical Nutrition*, 2006, 60, 567-72.

77. Abrams S.A., Griffin I.J., Hawthorne K.M., Ellis K.J. Effect of prebiotic supplementation and calcium intake on body mass index. *The Journal of Pediatrics*, 2007, 151, 293-8.

78. Fiordaliso M., Kok N., Desager J.P., Goethals F., Deboyser D., Roberfroid M., Delzenne N. Dietary oligofructose lowers triglycerides, phospholipids and cholesterol in serum and very low density lipoproteins of rats. *Lipids*, 1995, 30 (2), 163-7.

79. Delzenne N.M., Kok N., Fiordaliso M..F., Deboyser D.M., Goethals F.M., Roberfroid M.B. Dietary fructooligosaccharides modify lipid metabolism in rats. *American Journal of Clinical Nutrition*, 1993, 57, 820S.

80. Kok N., Roberfroid M., Delzenne N. Dietary oligofructose modifies the impact of fructose on hepatic triacylglycerol metabolism. *Metabolism*, 1996, 45 (12), 1547-50.

81. Trautwein E.A., Rieckhoff D., Erbersdobler H.F. Dietary inulin lowers plasma cholesterol and triacylglycerol and alters biliary bile acid profile in hamsters. *The Journal of Nutrition*, 1998, 128, 1937-43.

82. Kok N., Roberfroid M.R., Robert A., Delzenne N. Involvement of lipogenesis in the lower VLDL secretion induced by oligofructose in rats. *British Journal of Nutrition*, 1996, 76, 881-90.

83. Kok N., Morgan L.M., Williams C.M., Roberfroid M.B., Thissen J-P., Delzenne M. Insulin, glucagon-like peptide 1, glucose-dependent insulinotropic polypeptide and insulin-like growth factor I as putative mediators of the hypolipidemic effect of oligofructose in rats. *The Journal of Nutrition*, 1998, 128, 1099-1103.

84. Davidson M.H., Maki K.C., Synecki C., Torri S.A., Drennan K.B. Effects of dietary inulin on serum lipids in men and women with hypercholesterolemia. *Nutrition Research*. 1998, 18 (3), 503-17.

85. Jackson K.G., Taylor G.R.J., Clohessy A.M., Williams C.M. The effect of the daily intake of inulin on fasting lipid, insulin and glucose concentrations in middle-aged men and women. *British Journal of Nutrition*, 1999, 82, 23-30.

86. Letexier D., Diraison F., Beylot M. Addition of inulin to a moderately high-carbohydrate diet reduces hepatic lipogenesis and plasma triacyglycerol concentrations in humans. *American Journal of Clinical Nutrition*, 2003, 77, 559-64.

87. Brighenti F. Dietary fructans and serum triacylglycerols: a meta-analysis of randomized controlled trials. *The Journal of Nutrition*, 2007, 137, 2552S-6S.

88. Kok N.N., Taper H.S., Delzenne N.M. Oligofructose modulates lipid metabolism alterations induced by a fat-rich diet in rats. *Journal of Applied Toxicology*, 1998, 18 (1), 47-53.

89. Daubioul C.A., Taper H.S., De Wispelaere L.D., Delzenne N.M. Dietary oligofructose lessens hepatic steatosis, but does not prevent hypertriglyceridemia in obese Zucker rats. *The Journal of Nutrition*, 2000, 130, 1314-9.

90. Rault-Nania M-H., Gueux E., Demougeot C., Demigné C., Rock E., Mazur A. Inulin attenuates atherosclerosis in apolipoprotein E-deficient mice. *British Journal of Nutrition*, 2006, 96, 840-4.

91. Reddy B.S., Hamid R., Rao C.V. Effect of dietary oligofructose and inulin on colonic preneoplastic aberrant crypt foci inhibition. *Carcinogenesis*, 1997, 18 (7), 1371-4.

92. Verghese M., Rao D.R., Chawan C.B., Williams L.L., Shackelford L.A. Dietary inulin suppresses azoxymethane-induced preneoplastic aberrant crypt foci in mature Fisher 344 rats. *The Journal of Nutrition*, 2002, 132, 2804-8.

93. Rowland I.R., Rumney C.J., Coutts J.T., Lievense L. Effect of *Bifidobacterium longum* and inulin on gut bacterial metabolism and carcinogen induced aberrant crypt foci in rats. *Carcinogenesis*, 1998, 2, 281-5.

94. Verghese M., Walker L.T., Shackelford L., Chawan C.B. Inhibitory effects of nondigestible carbohydrates of different chain lengths on azoxymethane-induced aberrant crypt foci in Fisher 344 rats. *Nutrition Research*, 2005, 25, 859-68.

95. Rafter J., Bennett M., Caderni G., Clune Y., Hughes R., Karlsson P.C., Klinder A., O'Riordan M., O'Sullivan G.C., Pool-Zobel B., Rechkemmer G., Roller M., Rowland I., Salvadori M., Thijs H., Van Loo J., Watzl B., Collins J.K. Dietary synbiotics reduce cancer risk factors in polypectomized and colon cancer patients. *American Journal of Clinical Nutrition*, 2007, 85, 488-96.

96. Hoebregs H. Fructans in foods and food products, ion-exchange chromatographic method: Collaborative study. *Journal of AOAC International*, 1997, 80 (5), 1029-37.

1.2 GALACTO-OLIGOSACCHARIDES

A. Nauta and H.C. Schoterman
Friesland Campina
Hanzeplein 25
8017 JD Zwolle
The Netherlands

1.2.1   Description

Nowadays, it is a well-established fact that the composition and activity of the microbial population (microbiota) in the gastro-intestinal tract (GIT) significantly contribute to the health and well-being of the host (1). As a result, there is a growing interest in functional foods and/or food ingredients that have a beneficial effect on human health and well-being through said microbiota. This is clearly illustrated by the increasing use of prebiotics: 'nondigestible food ingredients that beneficially affect the host by selectively stimulating the growth and/or activity of one or a limited number of bacteria in the colon and thus improve health' (2). Several classes of prebiotic oligosaccharides have been described of which galacto-oligosaccharides (GOS), fructo-oligosaccharides (FOS) and inulin are the most common. GOS have attracted particular attention because they are similar to the human (breast)milk oligosaccharides (HMOs). HMOs modulate the microbiota in the gut, affecting different gastrointestinal activities, and have the potential to influence inflammatory and immunological processes (3). GOS produced by enzymes (also called transgalacto-oligosaccharides, transgalactosylated oligosaccharides, trans-GOS, TOS or oligogalactosyl-lactose) have been shown to have similar beneficial effects as HMOs (4).

1.2.2   Manufacturing

One of the most commonly produced prebiotic oligosaccharides are GOS. They can be obtained by the enzymatic conversion of lactose (milk sugar) with the enzyme β-galactosidase. Lactose is a disaccharide that consists of β-D-galactose and β-D-glucose bonded through a β1-4 glycosidic linkage, and is usually purified from cow's milk whey (5). The β-galactosidase can mediate both the hydrolysis and polymerisation of β-linked sugars. Under the specific conditions used in the commercial GOS production process, however, the enzyme reacts with the available lactose to form an oligosaccharide, liberating a glucose molecule. The consecutive trans-galactosylation reactions with lactose or the formed oligosaccharides of different chain length as a donor, give rise to heterogeneous mixtures of (β-linked, β-) GOS with varying chain length and linkages (Figs.1.2.1 and 1.2.2). The amount and type of GOS produced depends on several factors, such as the enzyme, lactose concentration and source, type of process and process conditions.

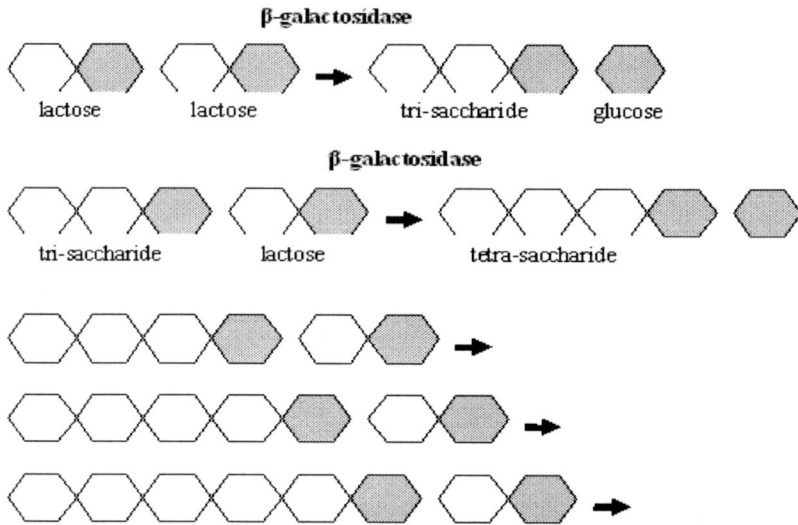

Fig. 1.2.1. Schematic representation of the consecutive trans-galactosylation reactions that give rise to the formation of the GOS mixture. Gal: galactose, Glc: glucose.

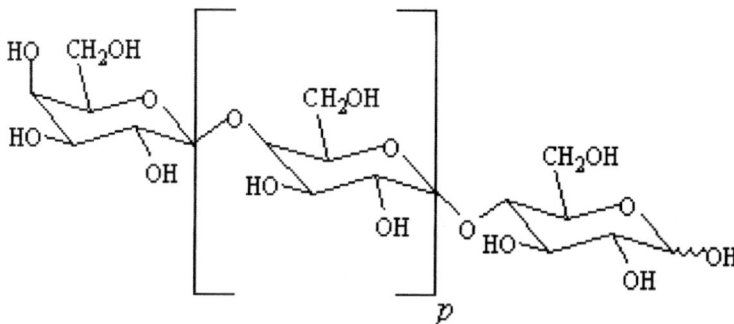

Fig. 1.2.2 General structure of GOS: a chain of variable numbers of galactose units, with a lactose moiety at the reducing end (P=0-6).

1.2.3   General properties

1.2.3.1   Composition

Although in principle, almost all glycosidic linkages can be formed during the production of GOS, β(1-4) and β(1-6) are the most abundant, linkages including 1→2 Glc, 1→3 Glc, 1→4 Glc, 1→6 Glc, 1→2 Gal and 1→3 Gal can also be present (6, 7, 8). The trisaccharides β(1-4)-galactosyl-lactose (4'-galacto-oligosaccharide) and β(1-6)-galactosyl-lactose (6'-galacto-oligosaccharide) present in commercial products are also found in human milk (9, 10). There are more similarities between commercially available GOS and oligosaccharides that occur in human milk. Like human milk oligosaccharides, commercial GOS contain a high amount of galactose and carry lactose at their reducing end (10). The saccharides in the commercially available Vivinal®GOS, available in syrup and powder form for wet and dry blending (FrieslandCampina DOMO, Zwolle, the Netherlands), vary in chain length from disaccharides (DP2) to octasaccharides (DP8).

1.2.3.2   Physicochemical characteristics

1.2.3.2.1 Heat and acid stability

GOS are extremely stable under pasteurisation and sterilisation conditions and in acid environments. Figure 1.2.3 overleaf shows the results of two experiments in which the stability of GOS (Vivinal®GOS) and two types of FOS were determined. No noticeable breakdown of GOS occurs at high temperatures (85 °C for 5 minutes) and in acidic conditions (up to pH 2), indicating that pasteurisation does not affect GOS. FOS and inulin were shown to be less stable in the conditions tested. GOS was also stable after subsequent storage at 30 °C for 7 days (Fig. 2B). Similar results were obtained for pasteurised GOS after storage in a soft-drink model (pH 3) for 6 months.

1.2.3.2.2 Tolerance

Fermentation of oligosaccharides at high dosages in the colon can result in the production of gases (such as hydrogen), that may lead to certain gastrointestinal tract (GIT) complaints in sensitive subjects. Consumption of GOS up to 20 g of GOS as part of a daily diet, however, is well tolerated by humans (11-13). This high dosage of GOS (Vivinal®GOS) has also been shown to result in a 34% lower excretion of hydrogen (produced upon fermentation of the oligosaccharide) in the breath as compared to FOS at a similar dosage. In addition, there were fewer complaints of flatulence after consumption of GOS(11).

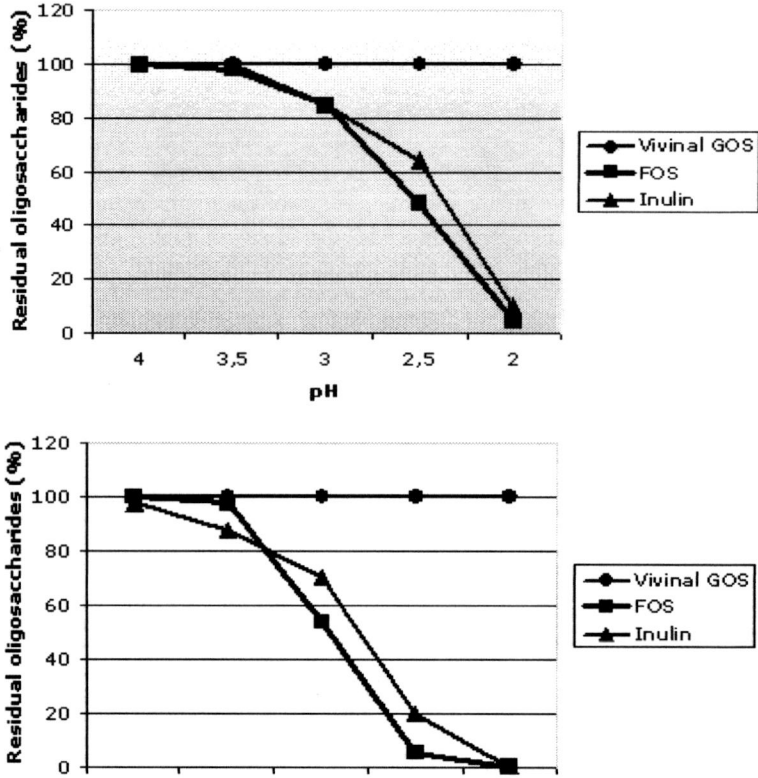

Fig. 1.2.3. **A.** *Stability of inulin, FOS and Vivinal®GOS after heating for 5 minutes at 85 °C at various pH levels.* **B.** *Results obtained for the same samples after subsequent storage for 7 days at 30°C.*

1.2.3.2.3 Caloric value

As GOS are not hydrolysed and/or absorbed in the small intestine, they can be considered to be low calorie carbohydrates. Only small amounts of energy are released through their fermentation in the colon. The caloric value of non-digestible oligosaccharides such as GOS, FOS and lactulose has been estimated to be 1-2 kcal/g (14, 15).The caloric value of GOS has been estimated to be 1.73 kcal/g (12), which is low in comparison with glucose (4 kcal/g).

1.2.4 Applications

The use of GOS is gradually increasing in various applications worldwide. GOS have a safe history of use in food and infant nutrition and are applied in various kinds of products. Products containing GOS were first launched in Japan in the late 1980s. In Europe, the first product containing GOS (a Dutch fermented milk product) was launched in 1997. Because of their high solubility and heat and acid stability, GOS are particularly suitable for use in acid products such as fruit juices and soft drinks and heat-treated items such as bakery products. GOS is increasingly incorporated into synbiotic formulations, which consist of both pre- and probiotics. Many probiotics are applied for their demonstrated health benefits, such as anti-pathogen activity and immune stimulation in the GIT. GOS is used to enhance the ability to survive, colonisation and/or functionality of these probiotics. More specialised applications include their use in infant nutrition, functional foods and clinical nutrition. The dosage of GOS varies per product. Infant nutrition contains up to 0.8 g GOS/100 ml product, while current functional foods contain up to 5 g GOS per 100 g food. In the USA, Vivinal®GOS has Generally Recognized As Safe (GRAS) status.

Since the early 1990s, GOS have been increasingly applied as ingredients for infant formula to mimic the biological functions of human oligosaccharides. For more than a decade, over 90% of the infant formulas in Japan have been supplemented with non-digestible oligosaccharides (NDOs) as growth-promoting factors for bifidobacteria (16). In Europe, mainly GOS or a combination of GOS/long-chain (lc)FOS (90% GOS and 10% lcFOS) are applied in infant formulas, follow-on formulas and growing-up milks. According to EU Directive 2006/141/EC on Infant Formulae and Follow-on Formulae, GOS can also be added to infant nutrition in all member states of the EU (17). In the USA, Vivinal®GOS has GRAS status for use in term infant formulas at a maximum concentration of 0.5 g/100 ml.

1.2.5   Physiological properties

1.2.5.1   Gut health

GOS are highly resistant to digestion and absorption during transit through the small intestine (14, 18, 19). The salivary and digestive enzymes and the acidic conditions of the stomach have virtually no effect on the acid-stable GOS. Upon consumption, these NDOs reach the colon fairly intact and are completely fermented by the health-promoting members of the gut microbiota.

After birth, the human GIT exists in symbiosis with the intestinal microbiota, composed of a large number and variety of bacteria. A healthy microbiota constitutes one that is predominantly carbohydrate-fermenting (saccharolytic), and comprises significant numbers of bifidobacteria and lactobacilli. Both species have been linked with increased resistance to infections and diarrhoeal disease, stimulation of immune system activity, protection against colon cancer, and the synthesis of various vitamins. Increasing evidence also suggests that the pathogenesis of GIT diseases and atopy is associated with an imbalance in the intestinal microbiota with a relative predominance of pathogenic bacteria, and paucity of the beneficial organisms.

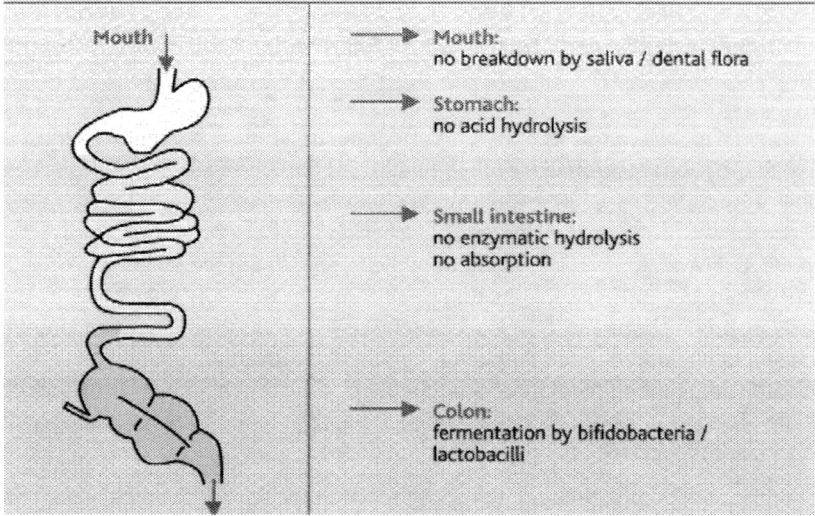

Fig. 1.2.4. The digestibility of GOS.

GOS have been shown to positively influence both the composition and activity of the microbiota. Through their effect on the microbiota, GOS also affect the activity of the immune system. In addition, GOS have several other effects that positively influence the health and well-being of the host. The various beneficial effects are discussed below; an overview is shown in Figure 1.2.5.

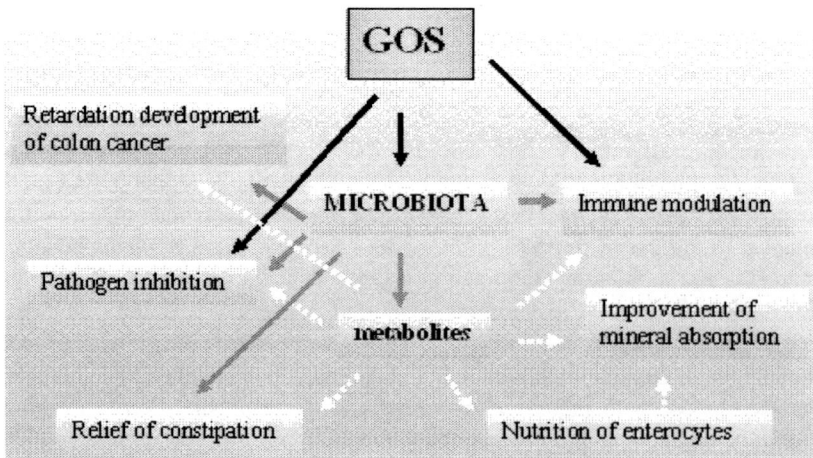

Fig. 1.2.5. Diagram showing the various beneficial effects of GOS. Adapted from (20).

1.2.5.2   Bifidogenic activity

Many studies with healthy subjects demonstrated increased numbers of bifidobacteria and/or lactobacilli in the faeces after the consumption of GOS (21-28). The bifodobacteria-stimulating (bifidogenic) activity has also been clearly demonstrated with GOS-supplemented and GOS/lcFOS-supplemented (90% GOS and 10% lcFOS) formulas in term and pre-term infants (6, 29-35).

Various clinical trials with GOS and GOS/lcFOS-supplemented infant formula for term infants have been published. In general, the supplementation was shown to elicit a dose-related bifidogenic response and increase in bifidobacterial predominance (29-31, 34). The microbial diversity and composition of the microbiota of GOS/lcFOS-supplemented infants was shown to closely resemble that of breast-fed infants, also at the level of the different Bifidobacterium species (35-37). In contrast, standard formula-fed groups of infants harbour a more adult-like microbiota. At the end of a 6-week study, it was found that the bifidobacteria and lactobacilli accounted for 80% of the faecal microbiota in breast-fed and GOS/lcFOS-supplemented groups, while only accounting for 50% in the standard formula group. The supplementation also gives rise to stool characteristics such as pH, short chain fatty acid (SCFA) profiles and consistency that more resemble those of breast-fed infants (35, 38). Administration of infant formulas containing GOS/lcFOS to pre-term infants gave similar results (7) with increased faecal bifidobacteria and softer stool consistency.

GOS have also been shown to have a synergistic effect on the bifidogenic activity of probiotics (39). The increase in the amount of bifidobacteria in school-aged children was significantly greater after the ingestion of GOS combined with *Lactobacillus rhamnosus* GG (LGG), compared to the ingestion of only LGG.

1.2.5.2   Support of natural defences

1.2.5.2.1 Inhibition of pathogens

GOS are able to selectively manipulate the intestinal microbiota, which indirectly results in the displacement of less desirable members of the microbiota (40, 41). In addition, the metabolism of GOS by specific members of the microbiota results in the production of antagonistic agents (42, 43) and SCFAs. The latter reduce the luminal pH in the colon to levels below those at which the pathogens can effectively multiply. GOS have also been shown to have a more direct inhibitory effect on pathogens, as they competitively prevent bacterial adherence. GOS resemble the receptor sites coating the intestinal epithelial cells, to which pathogens adhere for initiation of the infection process (44). As a result, they can act as 'molecular receptor decoys' or 'anti-adhesives' that competitively inhibit bacterial adherence by mimicking the host cell receptors (44-46). GOS have been shown to impair the adherence of an enteropathogenic *Escherichia coli* (EPEC) strain on HEp-2 and Caco-2 cells by 65 and 70%, respectively, in a dose-dependent manner (46). In infants fed a GOS/lcFOS (90% GOS and 10% lcFOS)-

supplemented formula, the number of pathogens in faecal samples was lower compared with infants fed a standard formula (47).

1.2.5.2.2 Immune modulating effects, allergy and infections

The epithelial cells of the intestine are part of the gut-associated lymphoid tissue (GALT) and play a crucial role in signalling and mediating innate immune responses. They also produce essential signals for the induction of memory pathways of the adaptive immune system. The adaptive immune system consists of B cells that produce antibodies against proteins (humoral immunity), and T cells that remove antigens and viral infected cells (cellular immunity). The composition and/or the activity of the microbiota influence the maturation and modulation of the immune system activity (48). This is clearly illustrated in germ-free animals that are shown to have an immature and poorly developed immune system (49). The absence of a normal microbiota can also result in increased antigen transport across the gut mucosa (50). As the gut microbiota is established, the capacity of the GALT to produce IgA-secreting cells increases.

With its effect on the microbiota, GOS indirectly influence mucosal and systemic immune activity (51). In addition, the increased production of SCFAs by GOS fermentation contributes to the maintenance of a non-inflammatory environment in the intestine as several of these SCFAs have been shown to modulate immune responses (52-55). For a GOS/lcFOS mixture (90% GOS and 10% lcFOS) a (partially) microbiota-independent effect on the immune response has also been demonstrated (56). It was shown to modulate vaccine-specific delayed-type hypersensitivity (DTH) responses. It increased the proportion of faecal bifidobacteria and lactobacilli and enhanced DTH responses dose-dependently. FOS/inulin induced similar effects on the gut microbiota. However, FOS/inulin did not enhance DTH responses, indicating that an increase in the proportions of bifidobacteria and lactobacilli from FOS/inulin consumption is not sufficient for the observed immuno-modulatory effect of GOS.

Secretory immunoglobulin A (sIgA) plays an important role in the defence of the GIT. Formula-fed infants who lack the transfer of protective maternal sIgA from breast milk can benefit from strategies to support maturation of (humoral) immunity and endogenous production of sIgA (57). In an intervention study, infants fed on a formula supplemented with a GOS/lcFOS mixture showed a trend towards higher faecal sIgA levels compared with the standard formula-fed infants. Infants fed on a probiotic (*Bifidobacterium lactis* BB12) formula showed a highly variable faecal sIgA concentration with no statistically significant differences compared with the standard formula group. A recent double-blind, randomised, placebo-controlled study also demonstrated higher concentrations of faecal sIgA after consumption of GOS/lcFOS-supplemented infant formula (58).

There is increasing evidence of a protective effect of prebiotics against manifestations of allergy as well. In both eczema and food allergy, there is evidence of an inflammatory response in the GIT (59). Infants with early onset allergic disease are also at risk of other allergic manifestations, a phenomenon described as 'the allergic march'. Atopic dermatitis (AD) is usually the first manifestation of allergy during early infancy. Breastfeeding has been reported to lower the incidence of atopy-related disorders (60-62), an effect that was also shown for a GOS/lcFOS mixture (63). GOS/lcFOS supplementation reduced the cumulative incidence of AD in high-risk infants by altering immune development during the first six months of life (Fig. 1.2.6). The supplementation was shown to induce beneficial total serum antibody profiles (reduced IgE levels), specifically modulating the immune response towards food allergens, while leaving vaccination responses intact (64, 65). The observed protective effect against allergic manifestations lasted beyond the intervention period up to 2 years of age (66).

The GOS/lcFOS mixture also proved to have a protective effect against infections; it was shown to reduce the incidence of infectious episodes during the first 2 years of life (66). Infants who received supplemented (GOS/lcFOS) hypoallergenic formula had fewer episodes of physician-diagnosed overall and upper respiratory tract infections. A study with infant formula supplemented with the probiotic *Bifidobacterium longum* BL999 and GOS/lcFOS also showed that children receiving this synbiotic treatment had a non-significant tendency toward fewer airway infections compared with the control group (67). Dietary supplementation of a combination of 4 probiotic strains and GOS on allergic diseases in allergy-prone infants significantly reduced eczema and IgE-associated eczema (68).

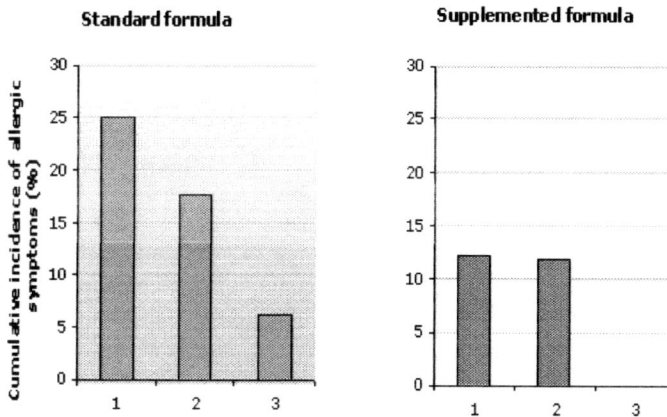

Fig. 1.2.6. Incidence of allergic symptoms in infants fed a standard formula and (GOS/lcFOS-) supplemented formula. 1, atopic dermatitis; 2, bronchial symptoms; 3, acute allergic cutaneous reactions. Source: (63).

1.2.5.2.3 Stimulation of the absorption of minerals

The dietary intake or bioavailability of minerals that play an important role in physiological processes is not always sufficient to meet the body's requirements, particularly in certain target groups. Several studies have shown that GOS stimulate the absorption of various minerals (26, 69-71). Most studied is the absorption of calcium, as it is required as a structural component of bones, and also plays an important role in blood coagulation and muscle contraction. The effect of GOS (Vivinal®GOS) on calcium absorption was demonstrated in postmenopausal woman in a double-blind randomised cross-over study (13). The consumption of a GOS-supplemented yoghurt drink did not result in elevated urinary calcium excretion, indicating that GOS also increase the uptake of calcium by the bones and/or inhibit bone resorption. This effect on bone was shown in a study with rats that were given a diet containing 5% (w/v) GOS for a period of 30 days (26, 69, 70). In addition to the increased absorption of calcium, GOS were shown to result in higher bone ash weight and calcium content in femur and tibia, indicative of the prevention of bone mineral loss (26). The bioavailability of magnesium and phosphorous is also positively influenced by GOS, as was demonstrated in magnesium-deficient rats (72).

Several mechanisms for the stimulation of mineral absorption by GOS fermentation have been proposed (73). For example, the SCFAs produced result in a reduction in pH levels, which can lead to an increase in the solubility and thus the absorption of minerals (69, 71). The SCFAs lactate and butyrate also promote the proliferation of enterocytes. The resulting enlargement of the absorption surface could also positively influence mineral absorption (73).

1.2.5.2.4 Alleviation of constipation

Several studies have demonstrated that the consumption of GOS can alleviate constipation in people who are constipated or who have a predisposition to this condition. For example, the defecation of healthy adults with a tendency for constipation significantly improved as manifested by an increased stool frequency and softer faeces (74). Other studies demonstrated similar beneficial effects in elderly subjects suffering from constipation (75, 76). A study with infant formula supplemented with the probiotic *Bifidobacterium longum* BL999 and a GOS/lcFOS mixture showed that children receiving this synbiotic treatment had less constipation than the control group (67). A number of mechanisms are thought to be involved in the improvement of bowel movement by GOS consumption, e.g. the stimulation of bacterial growth could result in an increase in bacterial biomass and faecal weight (14). The SCFAs that are subsequently produced could also stimulate intestinal peristalsis and increase faecal moisture with osmotic pressure (77).

1.2.5.2.5 Retardation of the development of colon cancer

Several fermentation-related processes that are associated with an increased colon cancer risk can be retarded by GOS consumption. The formation of secondary bile acids is inhibited by the reduction in the colonic pH as a result of GOS fermentation (78). GOS also give rise to a reduction in the activity of several genotoxic bacterial enzymes involved in the formation of toxic and carcinogenic compounds (79). Ammonia, indoles and the amino acid metabolites phenol and p-cresol have also been linked to the development of colon cancer (11). In a study with healthy subjects, a decrease in the concentration of ammonia, p-cresol and indoles in the faeces was observed after the consumption of GOS (20, 80). GOS also suppress the production of phenols in the intestinal tract and the accumulation of phenols in the serum (81). GOS has been shown to be highly protective against the development of induced colorectal tumours in rats (82, 83). Rats were fed diets with either a low or a high dose of GOS and a low, medium or high amount of fat. A high dosage of GOS resulted in a significant reduction in the multiplicity of adenomas, carcinomas and the total number of tumours. The incidence and size of the tumours were also reduced, irrespective of the amount of fat in the rats' diet.

1.2.6   Analytical method

For a quantitative analysis of GOS in food products, an AOAC-approved HPAEC-PAD method (High Performance Anion Exchange Chromatography, coupled with Pulsed Amperometric Detection) is used (84). This method gives rise to a high-resolution separation and selective detection of carbohydrates in (complex) food matrices. The GOS are extracted from a food sample and subsequently subjected to the action of a β-galactosidase. The enzyme hydrolyses all terminal, non-reducing β-D-galactose residues present in the β-D-galactosides such as GOS and lactose. The lactose and galactose concentrations in the enzyme-treated and reference (untreated) sample are then determined by HPAEC-PAD. The concentration of GOS can be calculated from both results.

1.2.7   References

1. Editorial, Who are we? *Nature*, 2008, 453, 563.

2. Gibson G.R., Roberfroid M.B. Dietary modulation of the human colonic microbiota: introducing the concept of prebiotics. *Journal of Nutrition*, 1995, 125, 1401-12.

3. Sharon N., Ofek I. Safe as mother's milk: carbohydrates as future anti-adhesion drugs for bacterial diseases. *Glycoconjugate Journal*, 2000, 17 (7-9), 659-64.

4. Boehm G., Stahl B. Oligosaccharides, in *Functional Dairy Products*. Eds Mattila T., Saarela, M. Cambridge, England, Woodhead Publishing Limited, 2003, 203-43.

5. Matsumoto K. Oligosaccharides, production, properties and applications, in Galactooligosaccharides, in *Japanese Technology Reviews*. Ed. Nakakuki T., Tokyo, Gordon and Breach Science Publishers, 1993, 3 (2), 90-106.

6. Ekhart P.F., Timmermans E. Techniques for the production of transgalactosylated oligosaccharides (TOS). *Bulletin of the IDF*. 1996. 313, 59-64.

7. Chockchaisawasdee S., Athanasopoulos V.I., Niranjan K., Rastall R.A. Synthesis of galacto-oligosaccharide from lactose using beta-galactosidase from Kluyveromyces lactis: Studies on batch and continuous UF membrane-fitted bioreactors. *Biotechnology. Bioengineering*, 2005, 89 (4), 434-43.

8. Schoterman H.C., Timmermans H.J.A.R. Galacto-oligosaccharides, in *Prebiotics and Probiotics, LFRA Ingredients Handbook*. Eds, Gibson, G.R., Angus, F., Leatherhead, Leatherhead Food RA Publishing, 2000, 19-46.

9. Yamashita K., Kobata A. Oligosaccharides of human milk. *Archives of Biochemistry and Biophysics*, 1974, 161, 164-70.

10. Boehm G., Lidestri M., Casetta P., Jelinek J., Negretti F., Stahl B., Marini A. Supplementation of a bovine milk formula with an oligosaccharide mixture increases counts of fecal bifidobacteria in preterm infants. *Archives of Disease in Childhood- Fetal Neonatal Edition*, 2002, 86, F178-F81.

11. Alles M.S., Schoterman H.C. Effects of fructo-oligosaccharides and galacto-oligosaccharides in breath hydrogen excretion and gastrointestinal wellbeing, *unpublished data*, 1999.

12. Van Dokkum W. Tolerantie voor galacto-oligosaccharide bij de mens, TNO Nutrition and Food Research Institute, The Netherlands, *Confidential report*, 1995.

13. Van den Heuvel E.G.H.M., Schoterman M.H.C., Muijs, T. Transgalactooligosaccharides stimulate calcium absorption in postmenopausal women. *The Journal of Nutrition*, 2000, 130, 2938-42.

14. Sako T., Matsumoto K., Tanaka R. Recent progress on research and applications of non-digestible galacto-oligosaccharides. *International Dairy Journal*, 1999, 9, 69-80.

15. Salminen S., Bouley C., Boutron-Runault M-C., Cummings J.H., Frank A., Gibson G.R., Isolauri E., Moreau M-C., Roberfroid M.,Rowland, I., Functional food science and gastrointestinal physiology and function. *British Journal of Nutrition*, 1998, 80 (S1), S147-71.

16. Boehm G., Stahl B., Jelinek J., Knol J., Miniello V., Moro G.E. Prebiotic carbohydrates in human milk and formulas. *Acta. Paediatrica, Suppl.* 2005, 94 (449), 18-21.

17. The Commission of the European Communities. Commission Directive 2006/141/EC of 22 December 2006 on infant formulae and follow-on formulae and amending Directive 1999/21/EC. Off J EU, Dec. 30, 2006:L401/1-33.

18. Asp N.G. Hydrolysis of galacto-oligosaccharides by human intestinal enzymes and acid, *unpublished data*, 1994.

19. Tanaka R., Takayama H., Morotomi M., Kuroshima T., Ueyama, S., Matsumoto, K., Kuroda, A., Mutai, M. Effects of administration of TOS and bifidobacterium breve 4006 on the human fecal flora. *Bifidobacteria Microflora*, 1983,2, 17-24.

20. Ouwehand A.C., Derrien M., de Vos W., Tiihonen K., Rautonen N. Prebiotics and other microbial substrates for gut functionality. *Current Opinion in Biotechnology*, 2005, 16, 212-7.

21. Ito M., Deguchi Y., Matsumoto K., Kimura M., Onodera N., Yajima T. Influence of galactooligosaccharides on the human fecal microflora. *Journal of Nutritional Science and Vitaminology*, 1993, 39, 635-40.

22. Ito M., Deguchi Y., Miyamori A., Matsumoto K., Kikuchi H., Matsumoto K., Kobayashi Y., Yajima T., Kan T. Effects of administration of galactooligosaccharides on the human faecal microflora, stool weight and abdominal sensation. *Microbial Ecology in Health and Disease*, 1990, 3, 285-92.

23. Bouhnik Y., Flourie B., D'Agay-Abensour L., Pochart P., Gramet G., Durand M., Rambaud J.C. Administration of transgalacto-oligosaccharides increases fecal bifidobacteria and modifies colonic fermentation metabolism in healthy humans. *Journal of Nutrition*, 1997, 127, 444-8.

24. Ishikawa F., Takayama H., Suguri T., Matsumoto K., Ito M., Chonana O., Deguchi Y., Kikuci H., Watanuki M. Effects of β-1-4 linked galacto-oligosaccharides on human fecal microflora. *Bifidus Microflora*, 1995, 9, 5-18.

25. Tamai S., Nakamura Y., Ozawa O. Yamauchi K. Effects of galactooligosaccharides intake on human fecal flora and metabolites (in Japanese). *Oyo Toshitsu Kagaku*, 1994, 41, 343-8.

26. Chonan O., Matsumoto K., Watanuki M. Effects of galactooligosaccharides on calcium absorption and preventing bone loss in ovariectomized rats. *Bioscience, Biotechnology,and Biochemistry*, 1995, 59 (2), 236-9.

27. Bouhnik Y., Raskine L., Simoneau G., Vicaut E., Neut C., Flourié B., Brouns F., Bornet F.R. The capacity of nondigestible carbohydrates to stimulate fecal bifidobacteria in healthy humans: a double-blind, randomised, placebo-controlled, parallel-group, dose-response relation study. *American Journal of Clinical Nutrition*, 2004, 80,1658-64.

28. Macfarlane G.T., Steeds H., Macfarlane S. Bacterial metabolism and health related effects of galactooligosaccharides and other prebiotics. *Journal of Applied.Microbiology*, 2008, 104 (2), 305-44.

29. Ben X.M., Zhou X.Y., Zhao W.H., Yu W.L., Pan W., Zhang W.L., Wu S.M., Van Beusekom C.M., Schaafsma A. Supplementation of milk formula with galacto-oligosaccharides improves intestinal micro-flora and fermentation in term infants. *Chinese Medical Journal*, 2004,117, 927-31.

30. Sawatzki G., Marten B., Scholz-Ahrens K.E., Schrezenmeir J., Vigi V., Fanaro S., Bagna R., Fabris C., Bertino E., Arguelles F., Quintana L.P., Valsasina R., Zelenka R. Double-blind, randomized, multicenter trial on the effects of prebiotic galacto-oligosaccharides on the fecal flora and development of infants: First results, in *The wonders of whey...catch the power*, Proceedings of the 4th International Whey Conference Chicago, 161-169, 2005.

31. Napoli J.E.A.C., Brand-Miller J.C., Conway P. Bifidogenic effects of feeding infant formula containing galacto-oligosaccharides in healthy formula-fed infants. Proceedings of the Nutrition Society Australia, 2003, 27. *Asia Pacific Journal of Clinical Nutrition*, 2003, 12, S60.

32. Fanaro S., Jelinek J., Stahl B., Boehm G., Kock R., Vigi V. Acidic oligosaccharides from pectin hydrolysate as new component for infant formulae: Effect on intestinal flora, stool characteristics, and pH. *Journal of Pediatric Gastroenterology and Nutrition*, 2005, 41, 186-90.

33. Mihatsch W.A., Hoegel J., Pohlandt F. Prebiotic oligosaccharides reduce stool viscosity and accelerate gastrointestinal transport in preterm infants. *Acta Pediatrica*, 2006, 95, 843-8.

34. Moro G., Minoli I., Mosca M., Fanaro S., Jelinek J., Stahl B., Boehm, G. Dosage related bifidogenic effects of galacto and fructo-oligosaccharides in formula fed term infants. *Journal of Pediatric Gastroenterology and Nutrition,* 2002, 34, 291-5.

35. Bakker-Zierikzee A.M., Alles M., Knol J., Kok F.J., Tolboom J.J.M., Bindels J.G. Effects of infant formula containing a mixture of galacto- and fructo-oligosaccharides or viable *Bifidobacterium animalis* on the intestinal microflora during the first four months of life. *British Journal of Nutrition*, 2005, 94, 783-90.

36. Haarman M., Knol J. Quantitative real-time PCR assays to identify and quantify fecal *bifidobacterium* species in infants receiving a prebiotic infant formula. *Applied and Environmental Microbiology*. 2005, 71, 2318-24.

37. Haarman M., Knol J. Quantitative real-time PCR analysis of fecal *Lactobacillus* species in infants receiving a prebiotic infant formula. *Applied and Environmental Microbiology*, 2006,72, 2359-65.

38. Knol J., Scholtens P., Kafka C., Steenbakkers J., Gro S., Helm K., Klarczyk M., Schöpfer H., Böckle H.M., Wells J., Colon microflora in infants fed formula with galacto- and fructo-oligosaccharides: more like breast-fed infants. *Journal of Pediatric Gastroenterology and Nutrition*, 2005, 40 (1), 36-42.

39. Piirainen L., Kekkonen R.A., Kajander K., Ahlroos T., Tynkkynen S., Nevala R., Korpela R. In school-aged children a combination of galacto-oligosaccharides and *Lactobacillus* GG increases bifidobacteria more than *Lactobacillus* GG on its own. *Annals of Nutrition and Metabolism*, 2008, 52, 204-8.

40. Fooks L.J., Gibson, G.R. Probiotics as modulators of the gut flora. *British Journal of Nutrition*, 2002, 88. S1, S39-49.

41. Gibson G.R., McCartney A.L., Rastall R.A. Prebiotics and resistance to gastrointestinal infections. *British Journal of Nutrition*, 2005, 93, S1, S31–4.

42. Gibson G.R., Wang X. Regulatory effects of bifidobacteria on the growth of other colonic bacteria. *Journal of Applied Bacteriology*. 1994, 77, 412-20.

43. Anand S.K., Srinivasan R.A., Rao L.K. Antibacterial activity associated with *Bifidobacterium bifidum* – II. *Cultured Dairy Products Journal*, 1985, 21-3.

44. Kunz C., Rudloff S., Baier W., Klien N., Strobel S. Oligosaccharides in human milk: structural, functional, and metabolic aspects. *Annual Review of Nutrition*, 2000,20, 699-722.

45. Tzortis G., Goulas A.K., Gee J.M., Gibson G.R. A novel galactooligosaccharide mixture increases the Bifidobacterial population numbers in a continuous *in vitro* fermentation system and in the proximal colonic contents of pigs *in vivo*. *Journal of Nutrition*, 2005, 135, 1726-31.

46. Shoaf K., Mulvey G.L., Armstrong G.D., Hutkins R.W. Prebiotic galactooligosaccharides reduce adherence of enteropathogenic *Escherichia coli* to tissue culture cells. *Infection and Immunity*, 2006, 74, 6920-8.

47. Knol J., Boehm G., Lidestri M., Negretti F., Jelinek J., Agosti M., Stahl B., Marini A., Mosca F. Increase of fecal bifidobacteria due to dietary oligosaccharides induces a reduction of clinically relevant pathogen germs in the faeces of formula-fed preterm infants. *Acta Paediatrica.*, 2005b 94,449, 31-3.

48. Schley P.D., Field C.J. The immune-enhancing effects of dietary fibers and prebiotics. *British Journal of Nutrition*, 2002, 87, S221-30.

49. Norin E., Midtvedt T. Interactions of microflora associated characteristics of the host; non-immune function. *Microbial Ecology in Health and Disease*, 2000,11, 186-93.

50. Isolauri E. Probiotics in human disease. *American Journal of Clinical Nutrition*, 2001, 73 (6), 1142S.

51. Vos A., M'Rabet L., Stahl B., Boehm G., Garssen J. Immune-modulatory effects and potential working mechanisms of orally applied nondigestible carbohydrates. *Critical Reviews in Immunology*, 2007, 27, 97-140.

52. Inan M.S., Rasoulpour R.J., Yin L., Hubbard A.K., Rosenberg D.W., Giardina C. The luminal short-chain fatty acid butyrate modulates NF-kappaB activity in a human colonic epithelial cell line. *Gastroenterology*, 2000, 118, 724-34.

53. Yin L., Laevsky G., Giardina C. Butyrate suppression of colonocyte NF-kappa B activation and cellular proteasome activity. *Journal of Biology and Chemistry*, 2001, 276, 44641-6.

54. Cavaglieri C.R., Nishiyama A., Fernandes L.C., Curi R., Miles E.A., Calder P.C. Differential effects of short-chain fatty acids on proliferation and production of pro- and anti-inflammatory cytokines by cultured lymphocytes. *Life Science*, 2003, 73, 1683-90.

55. Tedelind S., Westberg F., Kjerrulf M., Vidal A. Anti-inflammatory properties of the short-chain fatty acids acetate and propionate: a study with relevance to inflammatory bowel disease. *World Journal of Gastroenterology*, 2007,13, 2826-32.

56. Vos A.P., Haarman M., Buco A., Govers M., Knol J., Garssen J., Stahl B., Boehm G., M'Rabet L. A specific prebiotic oligosaccharide mixture stimulates delayed-type hypersensitivity in a murine influenza vaccination model. *International. Immunopharmacology*, 2006, 6, 1277-86.

57. Bakker-Zierikzee A.M., van Tol E.A.F., Kroes H., Alles M.S., Kok F.J., Bindels J.G. Fecal SIgA secretion in infants fed on pre- or probiotic infant formula. *Pediatric.Allergy and Immunology*, 2006,17, 134-40.

58. Scholtens P.A., Alliet P., Raes M., Alles M.S., Kroes H., Boehm G., Knippels, L.M., Knol J., Vandenplas Y. Fecal secretory immunoglobulin A is increased in healthy infants who receive a formula with short-chain galacto-oligosaccharides and long-chain fructo-oligosaccharides. *Journal of Nutrition*, 2008, 138 (6), 1141-7.

59. Majamaa H., Isolauri E. Probiotics: a novel approach in the management of food allergy. *Journal of Allergy and Clinical Immunology*, 1997, 99, 179-85.

60. Gdalevich M., Mimouni D., David M., Mimouni M. Breast-feeding and the onset of atopic dermatitis in childhood: a systematic review and meta-analysis of prospective studies. *Journal of the American Academy of Dermatology*, 2001, 45 (4), 520-7.

61. Gdalevich M., Mimouni D., Mimouni M. Breast-feeding and the risk of bronchial asthma in childhood: a systematic review with meta-analysis of prospective studies. *Journal of Pediatrics*, 2001, 139 (2), 261-6.

62. Van Odijk J., Kull I., Borres M.P., Brandtzaeg P., Edberg U., Hanson L.A., Høst A., Kuitunen M., Olsen S.F., Skerfving S., Sundell J., Wille S. Breastfeeding and allergic disease: a multidisciplinary review of the literature (1966-2001) on the mode of early feeding in infancy and its impact on later atopic manifestations. *Allergy*, 2003, 58 (9), 833-43.

63. Moro G., Arslanoglu S., Stahl B., Jelinek J., Wahn U., Boehm G. A mixture of prebiotic oligosaccharides reduces the incidence of atopic dermatitis during the first six months of age. *Archives of Disease in Childhood*, 2006, 91, 814-9.

64. Garssen J., Moro G., Bruzzese E., Alliet P. A mixture of short chain galacto-oligosaccharides and long chain fructo-oligosaccharides induce san anti-allergic immunoglobulin profile in infants at risk, Poster PG4-15. Presented at ESPGHAN 2007, Barcelona, Spain, 9 - 12 May 2007.

65. Van Hoffen E., Ruiter B., Faber J., M'rabet L., Knol EF., Stahl B., Arslanoglu S., Moro G., Boehm G., Garssen J., A specific mixture of short-chain galacto-oligosaccharides and long-chain fructo-oligosaccharides induces a beneficial immunoglobulin profile in infants at high risk for allergy. *Allergy*, http://dx.doi.org/10.1111/j.1398-9995.2008.01765.x, 2008.

66. Arslanoglu S., Moro G.E., Schmitt J., Tandoi L., Rizzardi S., Boehm G. Early dietary intervention with a mixture of prebiotic oligosaccharides reduces the incidence of allergic manifestations and infections during the first two years of life. *Journal of Nutrition*, 2008, 138 (6), 1091-5.

67. Puccio G., Cajozzo C., Meli F., Rochat F., Grathwohl D., Steenhout P. Clinical evaluation of a new starter formula for infants containing live *Bifidobacterium longum* BL999 and prebiotics. *Nutrition*, 2007, 23 (1), 1-8.

68. Kukkonen K., Savilahti A., Haahtela T., Juntunen-Backman K., Korpela R., Poussa T., Tuure, T., Kuitunen M. Probiotics and prebiotic galacto-oligosaccharides in the prevention of allergic diseases: a randomized, double-blind, placebo-controlled trial. *Journal of Allergy and Clinical Immunology*, 2007, 119 (1), 192-8.

69. Chonan O., Watanuki M. Effect of galactooligosaccharides on calcium absorption in rats. *Journal of Nutritional Science and Vitaminology*, 1995, 41, 95-104.

70. Chonan O., Watanuki M. The effect of 6'-galactooligosaccharides on bone mineralization of rats adapted to different levels of dietary calcium. *International. Journal for Vitamin and Nutrition Research*, 1996. 66, 244-9.

71. Yanahira S., Morita M., Aoe S., Suguri T., Takada Y., Miura S., Nakajima I. Effects of lactitol-oligosaccharides on calcium and magnesium absorption in rats. *Journal of Nutritional Science and Vitaminology*, 1997, 43, 123-32.

72. Pérez-Conesa D., López G., Abellán P., Ros G. Bioavailability of calcium, magnesium and phosphorus in rats fed probiotic, prebiotic and symbiotic powder follow-up infant formulas and their effect on physiological and nutritional parameters. *Journal of the Science of Food and Agriculture*, 2006, 86 (14), 2327-36.

73. Scholz-Ahrens K.E., Ade P., Marten B., Weber P., Timm W., Acil Y., Glüer C.C., Schrezenmeir J. Prebiotics, probiotics, and synbiotics affect mineral absorption, bone mineral content, and bone structure. *Journal of Nutrition*, 2007, 137 (3), 838S-46S.

74. Deguchi Y., Matsumoto K., Ito A., Watanuki M. Effects of beta 1-4 galactooligosaccharides administration on defaecation of healthy volunteers with a tendency to constipation. *Japanese Journal of Nutrition*, 1997, 55 (1), 13-22.

75. Teuri U., Korpela R. Galacto-oligosaccharides relieve constipation in elderly people. *Annals of Nutrition and Metabolism*, 1998, 42, 319-27.

76. Shitara A. Effect of 4'-galactosyl-lactose on constipated old patients and intestinal bacteria. *Medical Biology*, 1988, 117, 371-3.

77. Tomomatsu H. Health effects of oligosaccharides. Ingestion of oligosaccharides increases the bifidobacteria population in the colon, which in turn contributes to human health in many ways. *Food Technology*, 1994, 48 (10), 61-5.

78. Ishikawa F., Takayama H., Suguri T., Matsumoto K., Ito M., Chonana O., Deguchi Y., Kikuci H., Watanuki M. Effects of β 1-4 linked galactooligosaccharides on human faecal microflora. *Bifidus Microflora*, 1995, 9, 5-18.

79. Rowland I.R., Tanaka R. The effects of transgalactosylated oligosaccharides on gut flora metabolism in rats associated with a human fecal microflora. *Journal of Applied Bacteriology*, 1993, 74, 667-74.

80. Ito M., Deguchi Y., Miyamori A., Matsumoto K., Kikuchi H., Matsumoto K., Kobayashi Y., Yajima T., Kan T. Effects of transgalactosylated disaccharides on the human intestinal microflora and their metabolism. *Journal of Nutritional Science and Vitaminology*, 1993, 39, 279-88.

81. Kawakami K., Makino I., Asahara T., Kato I., Onoue M. Dietary galacto-oligosaccharides mixture can suppress serum phenol and p-cresol levels in rats fed tyrosine diet. *Journal of Nutritional Science and Vitaminology*, 2005, 51(3), 182-6.

82. Wijnands M.V.W., Appel M.J., Hollanders V.M.H., Woutersen R.A. A comparison of the effects of dietary cellulose and fermentable galacto-oligosaccharide, in a rat model of colorectal carcinogenesis: fermentable fibre confers greater protection than non-fermentable fibre in both high and low fat backgrounds. *Carcinogenesis*, 1999, 20 (4), 651-6.

83. Wijnands M.V.W., Schoterman H.C., Bruijntjes J.B., Hollanders V.M., Woutersen, R.A. Effect of dietary galactooligosaccharides (GOS) on azoxymethane-induced colorectal cancer in Fischer 344 rats. *Carcinogenesis*, 2001, 22 (1), 127-32.

84. De Slegte J. Determination of trans-galactooligosaccharides in selected food products by ion-exchange chromatography: Collaborative study, *Journal of the Association of Analytical Chemists International*, 2002, 85, 417-23.

1.3 LACTULOSE (GALACTO-FRUCTOSE)

Pauline Quierzy
Solvay SA
25, rue de Clichy
F-75442 Paris Cedex 09
France

1.3.1   Description

Lactulose is otherwise known as galacto-fructose, and is a semi-synthetic disaccharide made from lactose as a result of an isomerisation, first described in 1929 by Montgomery and Hudson (1). Lactulose is a prebiotic that is commonly used in both the pharmaceutical industry and in food applications. Therefore, as this chapter covers the food applications only, the prebiotic will be referred to as galacto-fructose throughout. Although not present in nature, galacto-fructose occurs in heat-treated milk products as a result of the catalyst-free isomerisation of lactose (2).

In 1957, galacto-fructose was named by Dr Petuely as the "bifidogenic factor" although the nomenclature of pro- and prebiotics came into use only much later (3). Since the 1950s, galacto-fructose has a long history of use in humans as a prebiotic.

Galacto-fructose is the first recognised bifidogenic prebiotic, with strong supporting evidence of its effect in the human gastrointestinal tract. Its health properties make its use relevant for the food and health industry.

Galacto-fructose is a food ingredient that is also used, at higher dosages, in the pharmaceutical area, for the treatment of specific medical conditions (to alleviate constipation and to prevent and treat hepatic encephalopathy). In these applications, the denomination commonly used is lactulose. For information, lactulose and galacto-fructose are synonyms. For over 40 years, the pharmacodynamics of lactulose have made it an efficacious and very safe medicine.

From a legal perspective, galacto-fructose is recognised as a standard nutritional food ingredient (not "novel food"). The prebiotic and intestinal effects of galacto-fructose in nutrition have been reviewed and related health claims approved. Several health properties of galacto-fructose have opened up new lines of research, notably in the areas of mineral absorption and weight management.

1.3.2 General properties

Galacto-fructose is an isomerisation product of lactose. The two are identical regarding empirical formula: $C_{12}H_{22}O_{11}$ and molecular weight: 342.3 g/mol.

During the isomerisation, the glucose moiety of lactose is converted into fructose, resulting in a molecule 4-O-β-D-galactopyranosyl-D-fructose, whose generic name is galacto-fructose.

In contrast to its substrate lactose, galacto-fructose cannot be split by human intestinal enzymes at the β – 1, 4 glycosidic bond. It escapes hydrolysis and reaches the colon intact, where it is fermented by the intestinal microbiota.

The chemical structures of lactose and galacto-fructose are shown in Figure 1.3.1.

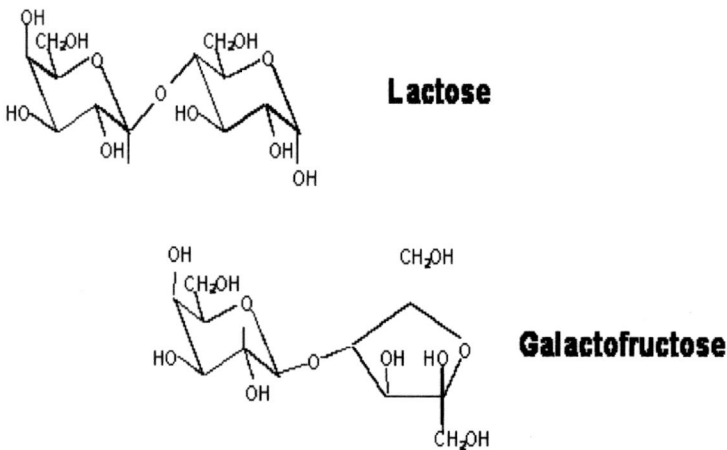

Figure 1.3.1 Structural formulae of lactose and galacto-fructose

Galacto-fructose is available in two different forms: as syrup liquid and as powder, obtained from the syrup after drying. Galacto-fructose syrup is a clear, colourless to pale brownish – yellow viscous liquid, miscible in water, its melting point is between 168.5 – 170.0 °C. Galacto-fructose powder is a white to creamish powder with high solubility and dispersibility in water (4). It is slightly soluble in methanol and insoluble in ether. Galacto-fructose has a sweet taste; the sweetness is 0.6 – 0.8 relative to sucrose (4).

Galacto-fructose syrup and powder are composed mainly of pure galacto-fructose and may contain a number of related residual sugars that result from the route of synthesis. These sugars are mainly lactose (the starting material for the product); epilactose and galactose may be present in smaller quantities (see Table 1.3.I).

TABLE 1.3.I
Principal composition of galacto-fructose, liquid and dry

Component	Chemical specifications (w/w)	
	Liquid (syrup) (for 100 g)	Dry (powder) (for 100 g)
Moisture	32 +/- 2 g	< 2 g
Dry substance	68 +/- 2 g	/
Galacto-fructose	50 +/- 2 g	74 +/- 2 g
Related sugars:		
Galactose	7 +/- 2 g	11 +/- 2 g
Lactose	3 +/- 2 g	6 +/- 2 g
Other carbohydrates (Epilactose, Tagatose, Fructose)	≤ 6 g	≤ 10 g

1.3.3   Applications

Galacto-fructose is a standard nutritional ingredient commercially available for the food industry in two different forms: syrup or powder. Galacto-fructose complies with all the quality requirements for food.

Both galacto-fructose powder and syrup are very resilient to heat and acidity. A series of tests with galacto-fructose have been performed using yogurts, biscuits, pancakes, chocolate, etc. to investigate its behaviour during processing, sensory properties, influence on taste, browning, and other properties (4). Major aspects are the flavour-enhancing properties; a slight browning behaviour without impact on the colouration of the food matrix at food dosage; excellent solubility in water; and a viscosity that allows substitution of sucrose (see Table 1.3.II).

TABLE 1.3.II
Selected chemical and physical properties of galacto-fructose

	Powder	Syrup
Taste improvement	√	√
Stabilising effect	√	√
Good solubility in water	√	NA
Viscosity closely resembles that of sucrose	NA	√
Relative sweetness 0.6-0.8 that of sucrose	√	√
Low energy value	1.1 kcal/g	0.7 kcal/g
Does not cause dental plaque	√	√

NA = non applicable

Galacto-fructose is highly adaptable to most food applications: biscuits and rusks, dietetics, cereals, food supplements, drinks and milk, yogurts, milk fermented products and fruit preparations.

In the production of biscuits, rusks and cereals, galactofructose offers particularly a good resistance to heat treatments. For drinks and milks, galacto-fructose offers a good resistance to acid pH, a high solubility and dispersibility. For most of these applications, incorporation of galacto-fructose does not modify texture neither viscosity. Its incorporation rate is low.

Incorporation of galacto-fructose in a food matrix is useful in specific applications, such as infant nutrition, clinical nutrition, sport nutrition and performance. In the field of infant nutrition, Dr Petuely, published core work as early as 1957 identifying galacto-fructose as a "bifidus factor" (3), which was confirmed by MacGillivray *et al.* (5). It was found that the colonic microflora of formula-fed babies was similar in composition to that of adults. However, if galactofructose was added to the formula milk, the infants had a faecal composition similar to that of breast-fed babies (see Table 1.3.III).

TABLE 1.3.III
Major components of colonic microflora of babies

Species of colonic microflora	Count per g faeces		
	Breast-fed babies	Formula-fed babies	Formula-fed + galacto-fructose
Bifidobacterium sp.	10^9	10^{10}	10^9
All anaerobes	10^9	10^{11}	10^9
All aerobes	10^{7-8}	10^8	10^{7-8}

At an early stage of the baby's life and in Asia and Japan especially, galacto-fructose was added to food products for babies. No other bifidogenic compounds were recognised at the time. Today, galacto-fructose is widely selected as one of the most adapted prebiotics for infant nutrition.

In clinical nutrition, supplementation with a small amount of galacto-fructose aims to improve health and well–being, by facilitating intestinal transit.

Finally, professional sportsmen frequently suffer from gastrointestinal symptoms due to high stress on these organs during competitions. Galacto-fructose helps prepare the digestive organs for times of high physical demand.

1.3.4   Physiological properties

Galacto-fructose has been extensively researched: its identity, efficacy, safety and security for consumption by humans are well documented. Many papers on its prebiotic action in particular have been published. While most of the claims made for prebiotics are based mainly on animal or *in vitro* trials, galacto-fructose data are based on clinical research to a considerable extent (6). The health properties of galacto-fructose extend far beyond its prebiotic effect.

1.3.4.1 Intestinal health

Galacto-fructose is a food ingredient with approved health claims for intestinal health. These claims have been validated by the Belgian Superior Council of Public Health:

"Galacto-fructose is a prebiotic bifidogene which helps the development of the intestinal flora and contributes to a better transit".

The recommended daily dose for such a health claim is 2.5 g galacto-fructose.

A complete scientific dossier has been submitted to the European National Authorities according to the procedure defined in Regulation (EC) 1924/2006 on nutrition and health claims,

1.3.4.1.1 The prebiotic effect

Lactulose is the original prebiotic, it was discovered in 1957, and its effects have been extensively researched .

The chemical structure of galacto-fructose explains its prebiotic effect. While lactose can be digested and absorbed by most of the white, Caucasian population, galacto-fructose cannot be split by human enzymes and is mainly metabolised by intestinal colonic bacteria. The biological potential of such an impact on the intestinal flora is widely and positively recognised.

Galacto-fructose passes through the stomach and ileum without any degradation, and reaches the colon intact. In the fasting state, galacto-fructose may reach the colon about one and a half hours after ingestion, as concluded from hydrogen breath test results. The colon represents a shelter for a huge population of bacteria, mainly composed of saccharolytic and proteolytic bacteria. The first group degrades carbohydrates by fermentation, whereas the second group breaks down proteins. Galacto-fructose is mainly metabolised by saccharolytic intestinal bacteria, including Bifidobacteria, Lactobacilli and, to a certain extent, Streptococci species. A significant increase in population levels of Bifidobacteria after supplementation with galacto-fructose characterises the prebiotic nature of this carbohydrate (7, 8). The promotion of such bacterial growth and activity significantly counteracts species such as Bacteroides, Clostridia, Salmonellae, Coliforms, and Eubacteria (10, 47). The proteolytic bacteria are considered as "unfriendly" for the organism because the nitrogen products resulting from their metabolism are potentially toxic.

The metabolisation of galacto-fructose results in a moderate rise in osmotic pressure and a clear and dose-dependent drop in intestinal pH. Physiologically, these two effects in combination lead to increased peristaltic action. With low dosages, which do not greatly decrease the colonic transit time, faecal pH is not affected. Major end products of bacterial metabolism are short-chain fatty acids (see Table 1.3.IV).

TABLE 1.3.IV
Concentration of volatile fatty acids in faeces of healthy volunteers after four weeks of supplementation with 2 x 10 g galacto-fructose or placebo (mmol/l) (9).

Acid	Placebo	Galacto-fructose
Lactic acid	14.7	19.6
Acetic acid	54.8	71.3
Propionic acid	13.0	11.1
Butyric acid	5.4	4.4

An overview of the mechanism is given in Figure 1.3.2.

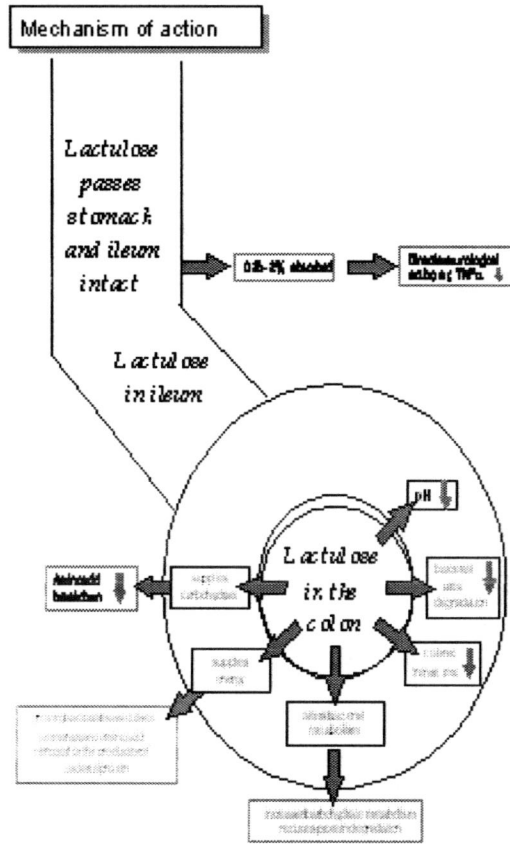

Figure 1.3.2 Principal mechanism of action of galacto-fructose

The consequences of feeding galacto-fructose on the indigenous flora of the intestinal tract are not shown in Figure 1.3.2. These are wide-ranging and a

summary overview is given in Table 1.3.V. As can be clearly seen from the
example of lactitol (11, 9) and from oligofructose and inulin (12), even very
closely related prebiotics may have significantly different effects on valuable
nutrients for intestinal bacteria. Consequently, the bacterial actions listed in effects
listed in Table 1.3.VI will vary for different prebiotics.

TABLE 1.3.V
Possible prebiotic effects of galacto-fructose

Acidification of gut contents
Ammonia depletion
Increased osmotic pressure
Increased peristalsis
Softening of stool
Facilitate defaecation
Selective bacterial growth
Stabilisation of the bacterial ecosystem
Inhibition of toxin-producing enzymes
Shorter residence time of toxins
Prevention of gastrointestinal infections (*Rotavirus*, *Candida*, etc.)
Prevention of carcinoma (colorectal and maybe other organs)
Prevention of enteritis
Prevention of inflammatory diseases
Anti-endotoxin effect (numerous possible applications)
Glucose and insulin control
Improved mineral absorption

The biological potential of galacto-fructose compares well with that of the effects
of other prebiotics, although clear differences do exist. For instance, lactitol has
the same empirical formula and molecular weight as galacto-fructose, and it is
also not digested in the small intestine, but it still shows very different effects on
several parameters of bacterial metabolism, such as faecal pH, and transit time (9,
11, 13, 14). Lactitol has been shown to be a significantly less effective prebiotic
at a similar dose. It has been suggested that only the cyclic part of this sugar-
alcohol acts as a prebiotic, while the alcohol residue is thought to act osmotically
(9). Direct comparisons of galacto-fructose and other prebiotics are scarce (15).

1.3.4.1.2 The transit effect

Historically, Mayerhofer and Petuely were the first to study the impact of galacto-
fructose on transit, and then suggest galacto-fructose for therapeutic use at high
dosage to relieve children from constipation (16). Due to the safety profile in
long–term use and the reliable efficacy of galacto-fructose, it has the best record
on safety for newborns and babies.

Galacto-fructose also facilitates more comfortable gut transit on healthy people
(17). This results from the combination of different effects, at different levels of
the intestinal tract. Before galacto-fructose is used by the intestinal bacteria, it

exerts an osmotic action in the small intestine (18). It draws water into the intestinal lumen, thereby increasing the mass of the stool. This in turn has a stimulating effect on the intestinal musculature and stimulates bowel movement. In addition, in the colon, total biomass, stool volume and osmotic pressure are increased and pH decreases, resulting in accelerated bowel movement and shorter transit time.

Regular daily consumption of galacto-fructose optimises functioning of the intestinal tract by regulating intestinal transit. Some factors in diet and lifestyle lead to small intestinal disorders, such as occasional constipation. Slow gut transit time causes discomfort and has been proposed as a promoting factor in the accumulation of toxic substances. Regulation of the intestinal transit time, especially in people with slow transit, improves overall well–being and quality of life.

There is a dose–related transit effect of galacto-fructose (19). Galacto-fructose is the only prebiotic that combines both a prebiotic and transit effect (20), at a daily dose of 2.5 g. Very few side effects are observed. The transit effect of galacto-fructose does not affect gastric emptying nor protein and lipid metabolism.

1.3.4.1.3 Mineral balance

Several clinical trials have shown the improvement in the intestinal absorption of calcium and magnesium after supplementation with galacto-fructose. This physiological effect was investigated particularly in postmenopausal healthy women (21) and, very recently, in adult men (22). The authors found a significant increase in calcium absorption without increased urinary excretion (3). The favourable effect of galacto-fructose on absorption of calcium, iron and zinc has been reviewed (23). These data confirmed results from animal experiments showing increased bone strength and reduced breakability after additional galacto-fructose when compared with inorganic calcium or whey (24, 25).

Recently, a trial initiated by the Solvay Pharmaceuticals Department (26) was completed, which measured the blood calcium and blood vitamin D concentrations after galacto-fructose or placebo in post-menopausal women with osteopenia but not osteoporosis. The authors found an increase of both calcium absorption and the blood vitamin D level.

1.3.5   Metabolic syndrome

Galacto-fructose combines a sweet taste with a low calorie intake, reduced by half with regard to the reference sugar. Moreover, supplementation with galacto-fructose leads to a moderate glycemic response. When compared with glucose, galactose, and lactose, galacto-fructose did not increase blood glucose levels in diabetic subjects. Genovese *et al.*, in a small cross-over trial, published preliminary results showing a statistically significantly reduced increase in blood glucose after oral glucose tolerance testing after ten days pre-treatment with galacto-fructose, when compared with no treatment (27). Further evidence for the antidiabetic activity of galacto-fructose was provided by Biancchi *et al.* (28). Hosaka *et al.* showed in the isolated jejunal loop of rats that galacto-fructose reduced the absorption of glucose by 40% without affecting amino acid uptake (29). Unpublished data (on file: International Lactulose Application Committee, Hannover, Germany) show that reduced glucose and insulin responses after galacto-fructose administration are not due to α-glucosidase-inhibiting properties. Cornell has suggested that endotoxins may reduce pancreatic insulin production and that endotoxin-reducing substances may have antidiabetic effects (30).

A recent study has also revealed a beneficial effect of galacto-fructose on insulin resistance. Resistance to insulin is a key factor in type 2 diabetes and cardiovascular diseases in obese patients, heart attack risk factors being part of the metabolic syndrome. The acetate produced from galacto-fructose colonic fermentation seems to be an antilipolytic agent that can reduce lipotoxicity and enhance insulin sensitivity (31).

Finally, scientific results support the idea that modulating gut microbiota could be beneficial for managing weight loss (32). Prebiotics and galacto-fructose in particular seem efficient at modifying gut microflora and possibly achieve these effects.

1.3.6   Immunity

1.3.6.1   Indirect immunological effects

Japanese workers have concentrated on the effects of bifidobacteria, and numerous specific and non-specific immunological effects exerted by bifidobacteria have been described. Since most bifidobacteria metabolise galacto-fructose well, the immunologic effects of bifidobacteria can most likely be triggered by feeding galacto-fructose (33).

1.3.6.2   Direct immunological effects

Although galacto-fructose is largely not absorbed and can be passed on to the liver and the systemic circulation, the small amounts that are absorbed (0.25–2%) are sufficient for immunological reactions (30 μg/ml). As it has been shown by Liehr *et al.* (34) and later by Greve *et al.* (35), galacto-fructose also shows specific immunological effects if given intravenously or *in vitro*. After a galactosamine challenge in rats, galacto-fructose almost completely prevented necrosis of hepatocytes and inflammatory reactions in liver tissue. Greve *et al.* found *in vitro* that the endotoxin-inactivating capacity of galacto-fructose was limited, and that the endotoxin-induced production of tumour necrosis factor by monocytes was significantly reduced (35). These results have been followed up by Solvay Pharmaceuticals. Preliminary results (unpublished) of *in vitro* trials show that macrophages activated by endotoxins of different origin (*Salmonella, E. coli, Pseudomonas*) showed differentiated reactions with regard to TNF-α, IL-1β, GM-CSF, IL-8, IL-2, TGF-β1, IL-10, IL-12, and IFN-γ. This suggests a direct immunological action of absorbed galacto-fructose in addition to indirect effects mediated by gut bacteria. Currently, these data do not allow any further conclusion, because the cytokine reactions require a differentiated evaluation.

1.3.6.3 Protective effect of galacto-fructose

In the colon, undigested galacto-fructose is an ideal nutritional basis for health as it promotes bacteria in the bowel flora. This can result in a large number of positive protective effects. Evidence has increased showing such a protective effect of galacto-fructose. The common basis of such a protection is seen in a shift of the bacterial composition and metabolism of the gut.

The differential growth in colonic bacteria is reflected by changes in enzymic activities with decreases of about 50–70%. A much reduced activity of some bacterial enzymes producing potentially toxic compounds (substances suggested as being co-carcinogens, for instance) has been observed (9).

Products of these enzymes and their potential effects are shown in Table 1.3.VI

.

TABLE 1.3.VI
Bacterial enzymes, products and potential effects

Enzyme	Product	Potential effects
Azoreductase	Aromatic amines	Carcinogenic/mutagenic
7α – dehydroxylase	Desoxycholic acid	Cell proliferation
B – glucuronidase	Aglycons	Carcinogenic/mutagenic
Nitroreductase	1 – aminopyrene	Mutagenic
Urease	Ammonia	Co-carcinogenic

Decreased production of other potential carcinogens such as ammonia may contribute to the protective effect of galacto-fructose; galacto-fructose inhibits bacterial ammonia production by acidifying the content of the colon and reducing the activity of the proteolytic bacteria. Moreover, the growing biomass uses ammonia and nitrogen from amino acids to synthesise bacterial protein. Galacto-fructose reduces colonic transit time, thus reducing the time available for ammonia production and expediting ammonia elimination.

The DNA-protective effect of galacto-fructose has been proposed (37, 38) and is still under investigation (39).

Much research is being done on the effects of butyric acid, which is a nutrient for colonocytes, reinforcing the intestinal barrier.

A potential preventive use of galacto-fructose in inflammatory bowel disease has been postulated (40), and the anti-inflammatory effects of galacto-fructose and prebiotics in general have been reviewed (41).

As early as 1980, an anti-endotoxin effect for galacto-fructose was suggested by Liehr *et al.* (42). Endotoxins (lipopolysaccharides, (LPS) from bacterial cell walls) trigger inflammatory reactions if they are translocated from the gut into the systemic circulation. Translocation can take place especially when the gut permeability is disturbed. This happens in all cases of trauma, e.g. burns, surgery, and even stress. In inflammatory bowel disease (Crohn's disease, ulcerative colitis) permeability is increased and may depend on the extent of the disease. Clinically, the anti-endotoxin effect of galacto-fructose was investigated: galacto-fructose supplementation for two or three days prior to an operation almost completely prevented endotoxin-dependent complications (43, 44). The anti-endotoxin effect of galacto-fructose may be better appreciated once the connection with metabolic disorders is better understood (45, 46).

Galacto-fructose enhances resistance to invading pathogens and thus prevents some infections to organisms such as *Candida* and *Salmonella* (47, 48).

1.3.7   Safety and tolerance

Galacto-fructose is very safe to use in food applications. Conn and Bircher confirmed that the indigestible carbohydrate or its biodegradation is not expected to produce any toxic intermediates (49).

Baskaran investigated the safety of galacto-fructose in growing rats with a view to assessing the safety of the addition of galacto-fructose to infant formula, and found very high calculated LD50 and maximum tolerated dosages, and neither mortality nor clinical signs of toxicity nor effects on food consumption have been found (50).

No putative or definite evidence of mutagenic, genotoxic or teratogenic effects of galacto-fructose has been obtained in human use. Animal studies have not revealed any teratogenic effects, and even high dosages have had no deleterious effects (51).

One aspect of feeding prebiotics is the bacterial production of gas. This may occasionally lead to meteorism and flatulence and seems to vary depending on the prebiotic (11, 15). Galacto-fructose supplementation, at low dosages in food applications, comes along with side effects similar to some common foods and compares well with similar prebiotic ingredients.

1.3.8   Analytical methods

Determination of galacto-fructose is mainly carried out in two settings: one is a diagnostic application of galacto-fructose used for the determination of gut permeability, and the other is measurement in processed dairy products, such as UHT-milk.

In permeability testing, galacto-fructose determination is usually done either by gas chromatography (52) or by means of simultaneous HPLC-determination of galacto-fructose and, for example, mannitol, although other or refined methods or determination from serum also are possible (53, 54).

Galacto-fructose occurs in UHT milk at concentrations of up to 0.5%. Methods for determination of galacto-fructose in milk products have been described by Villamiel *et al.* (15).

1.3.9 References

1. Montgomery E., Hudson C.S. Transformation of lactose to a new disaccharide, lactoketose. *Science*, 1929, 69, 556-7.

2. Geier H., Klostermeyer H. Formation of lactulose during heat treatment of milk. *Milchwissenschaft.* 1983, 38, 475-77.

3. Petuely F. Der Bifidusfaktor. *Deutsche Medizinische Wochenschrift*, 1958, 82, 1957-60.

4. Battermann W. Lactulose powder – main technological properties and its relevance in functional food. *International report Solvay Deutschland GmbH*, 1997; report n°AWTLAC 12.041.97.

5. MacGillivray P.C., Finlay H.V., Binns T.B. Use of lactulose to create a preponderance of lactobacilli in the intestine of bottle-fed infants. *Scottish Medical Journal*, 1959, 4, 182-9.

6. Huchzermeyer H., Schumann C. Lactulose, a multifaceted substance. *Gastroenterology*, 1997, 65, 945-55 .

7. Terada A., Hara H., Katapka M., Mitsuoka T. Effect of lactulose on the composition and metabolic activity of the human faecal flora. *Microbial Ecology in Health and Disease*, 1992, 5, 43-50.

8. Tomoda T. Nakano Y., Kageyama T. Effect of yoghurt and yoghurt supplemented with Bifidobacterium and/or Lactulose in healthy persons: a comparative study. *Bifidobacteria Microflora*, 1991, 10, 123.

9. Ballongue J., Schumann C., Quignon P. Effects of lactulose and lactitol on colonic microflora and enzymatic activity. *Scandinavian Journal of Gastroenterology*, 1997, 32 (S222), 41-4.

10. Tuohy K.M. Ziemer C.J., Klinder A. A human volunteer study to determine the prebiotic effects of lactulose powder on human colonic microbiota. *Microbial Ecology in Health and Disease*, 2002, 14, 165-73.

11. Kitler M. E. Luginbuhl M., Lang O., Wuhl F., Wyss A., Lebek, G. Lactitol and lactulose. An *in vivo* and *in vitro* comparison of their effect on the human intestinal flora. *Drug Investigation*, 1992, 4 (1), 73-82.

12. Wang X., Gibson G.R. Effects of the *in vitro* fermentation of oligofructose and inulin by bacteria growing in the human large intestine. *Journal of Applied Bacteriology*, 1993, 75, 373-80.

13. Ballongue J., Crociani J., Grill J.P. *In vitro* study of the effect of lactulose and lactitol on growth and metabolism of intestinal bacteria. *Gut*, 1995, 37 (S2), A48.

14. Pontes F.A., Silva A.T, Cruz A.C.,l. Colonic transit time and the effect of lactulose and lactitol in hospitalized patients. *European Journal of Gastroenterology and Hepatology*, 1995, 7, 441-5.

15. Villamiel M., Corzo N. Lactulose formation during batch microwave treatment of different types of dairy products. *Milchwissenschaft*, 1998, 53 (9), 487-90.

16. Mayerhofer F., Petuely F. Untersuchungen zur regulation der darmtragheit des erwachsenen mit hilfe der alctulose (Bifidus faktor). *Wein Klin Wochenschr*, 1959, 71, 865-9.

17. Bartram H.P., Scheppach W., Gerlach S., Ruckdeschel G., Kelber E., Kasper H. Does yogurt enriched with Bifidobacterium longum affect colonic microbiology and fecal metabolites in health subjects? *American Journal of Clinical Nutrition*, 1994, 59 92) 428-32.

18. Miller M.A., Parkman H. P., Urbain J-L. C., Brown K. L., Donahue D. J., Knight L. C., Maurer A. H., Fisher R. S. Comparison of scintigraphy and lactulose breath test hydrogen for assessment of orocaecal transit: lactulose accelerates small bowel transit. *Digestive Diseases and Sciences*. 1997, 42 (1), 10-18.

19. Bass P., Dennis S. The laxative effects of lactulose in normal and constipated subjects. *Journal of Clinical Gastroenterology*, 1981, 3 (1), 23-8.

20. Geboes K.P., Luypaerts A., Rutgeerts P., Verbeke K. Inulin is an ideal substrate for a hydrogen breath test to measure the orocaecal transit time. *Alimentary Pharmacology and Therapeutics*, 2003, 18 (7), 721-9.

21. Van den Heuval E.G.H.M., Mujis T., Van dokkum W., Schaafsma G. Lactulose stimulates calcium absorption on postmenopausal women. *Journal of Bone and Mineral Research*, 1999. 14. 1211-6.

22. Seki N., Hamano H., Ilyama Y., Asano Y., Kokubo S., Yamauchi K., Tamura Y., Uenishi K., Kudou H. Effect of lactulose on calcium and magnesium absorption: a study using stable isotopes in adult men. *Journal of Nutritional Science and Vitaminology*, 2007, 53 (1) 5-12.

23. Van den Heuvel E.G.H.M. Weidauer T. Role of the non-digestible carbohydrate lactulose in absorption of calcium. *Medical Science Monitor*, 1999, 5 (6), 1231-7.

24. Brommage R., Juillerat M.A. Jost R. Influence of Casein Phosphopeptides and lactulose on intestinal calcium absorption in adult female rats. *Le Lait*, 1991, 71 (2), 173-80.

25. Igarashi C., Ezawa I. Effects of whey calcium and lactulose on the strength of bone in ovariectomized osteoporosis model rats. *Pharmacometrics*, 1991, 42 (3), 245-54.

26. Solvay Pharmacueticals. Study to investigate if adding lactulose to vitamin D and calcium has an effect on the preservation of bone in postmenopausal women. http://clinicaltrials.gov/ct2/show/NCT00160264?intr=%22Lactulose%22&rank=7 2007. Clinical trial no. NCT00160264.

27. Genovese S., Riccardi G., Rivellese A.A. Lactulose improves blood glucose response to an oral glucose test in non-insulin dependent diabetic patients. *Diabetes, Nutrition and Matabolism*, 1993, 5 (4), 295-7.

28. Bianchi G.P., De Mitri M.S, Bugianesi E., Abbiati R., Fabbri A., Marchesini G. Lowering effects of a preparation containing fibres and lactulose on glucose and insulin levels in obesity. *Italian Journal of Gastroenterology*, 1994, 26, 174-8.

29. Hosaka H., Tokunaga K., Sakumoto I. The influence on intestinal absorption. *Gastroenterology Japan*, 1972, 7, 316-7.

30. Cornell R.P. Endogenous gut-derived bacterial endotoxin tonically primes pancreatic secretion of insulin in normal rats. *Diabetes*, 1985, 34, 1253- 9.

31. Ferchaud-Roucher V., Pouteau E., Piloquet H., Zaïr Y., Krempf M. Colonic fermentation from lactulose inhibits lipolysis in overweight subjects. *American Journal of Physiology Endocrinology, and Metabolism*, 2005, 289 (4), E716-20.

32. Turnbaugh P.J., Ley R. E., Mahowald M. A., Magrini V., Mardis E. R. Gordon J. I. An obesity-associated gut microbiome with increased capacity for energy harvest. *Nature*, 2006. 444, 1027-131.

33. Schumann C. Die immunologischen effekte der Lactulose. *Notabene medicine*, 1997, 27, 288-90.

34. Liehr H., Englisch G., Rasenack U. Lactulose – a drug with antiendotoxine effect. *Hepatogastroenterology*, 1980, 27, 356-60.

35. Greve J.W.N, Gouma D. J., von Leeuwen P.A., Buurman W. A. Lactulose inhibits endotoxin-induced tumour necrosis factor production by monocytes. An *in vitro* study. *Gut*, 1990, 31, 198-203.

36. Westphal G. Zur Ernährungsphysiologie und toxikologie von Maillard reaktionsprodukten.1. reaktionsmechanismen *in vitro* und *in vivo*. *Ernahrungs-Umschau*, 1992, 39. 450-2 .

37. Josimovic I.B. Die Verhütung des dickdarmkrebses mit Lactulose. *Krebsgeschehen*, 1979, 6, 175-8.

38. Rowland I.R., Bearne C.A., Fischer R., Pool-Zobel B.L. The effect of lactulose on DNA damage induced by DMH in the colon of human flora-associated rats. *Nutrition and Cancer*, 1996, 26, 37-47.

39. van Berge Henegouwen G.P., van der Werf S.D., Ruben A.T. Effect of long term lactulose ingestion on secondary bile salt metabolism in man: potential protective effect of lactulose in colonic carcinogenesis. *Gut*, 1987, 28, 675-80.

40. Liao W., Cui X- S., Jin X-. Y., Florén C-H. Lactulose, a potential drug for the treatment of inflammatory bowel disease. *Medical Hypotheses*, 1994, 43 (4), 234-8.

41. Kennedy R.J., Kirk S.J., Gardiner K.R. Promotion of favorable gut flora in inflammatory bowel disease. *Journal of Parenteral and Enteral Nutrition*, 2000, 24 (3), 189-95.

42. Liehr H., Englisch G., Rasenack U. Treatment of endotoxemia in galactosamine hepatitis by lactulose administered intravenously. *Hepato-Gastroenterology*, 1981, 28, 296-8.

43. Pain J.A., Cahill C.J., Gilbert J.M., Johnson C.D., Trapnell J. E., Bailey M.E. Prevention of postoperative dysfunction in patients with obstructive jaundice: a multicentre study of bile salts and lactulose. *British Journal of Surgery*, 1991, 78, 467-9.

44. Ozcelik M.F., Pekmezci S., Altinli E., Eroğlu C., Göksel S., Göksoy E. Lactulose to prevent translocation in biliary obstruction. *Digestive Surgery*, 1997, 14 (4), 267-71.

45. Tabatabaie T., Kotake Y., Wallis G., Jacob J.M., Floyd R.A. Spintrapping agent phenyl-N-tert-butylnitrone protects against the onset of drug-induced insulin-dependent diabetes mellitus. *FEBS*, 1997, 407, 148-52.

46. Yelich M.R., Schieber C.K., Umporowicz D.M., Filkins J.P. Polymixin-B suppresses endotoxin-induced insulin hypersecretion in pancreatic islets. *Circulatory Shock*, 1992, 38, (2), 85-90

47. Mack D.J., Smart L., Girdwood A., Scott P.J.W., Fulton J.D., Erwin L. Infection prophylaxis with Lactulose. *Age and Ageing*, 1993, 22 Supp (2), 8,18.

48. Tomada T., Nakano Y., Kageyama T. Intestinal Candida overgrowth and Candida infection in patients with leukemia: effect of Bifidobacterium administration. *Bifidobacteria Microflora*, 1988, 7, 71-4.

49. Conn H.O., Bircher J. Intestinal Candidia overgrowth and Candidia infection in patients with leukemia:effect of Bifidobacteria encephalopathy in *Hepatic encephalopathy: Syndromes and therapies*. Bloomington, Illinois, Medi-ed Press. 1994, 399-412.

50. Baskaran, V., Narasimhamurthy K., Nagendra R., Lokesh B.R. Safety evaluation of Lactulose syrup in rats. *Journal of Food Science and Technology*, 1999, 36, 355-357.

51. Baglioni J., Dubini F. Lattulosio:valutazione tossicological. *Bollettino chemico farmacentico*. 1976, 115, 596-606.

52. Laker M.F. Estimation of disaccharides in plasma and urine by gas liquid chromatography. *Journal of Chromatography*, 1979, 163, 9-18.

53. Marsilio R., D'Antiga L., Zancamn L., Dussini N., Zacchello F. Simultaneous HPLC determination with light scattering detection of lactulose and mannitol in studies of intestinal permeability in pediatrics. *Clinical Chemistry*, 1998, 44, 1685-91.

54. Cox M.A., Iqbal T.H., Cooper B.T., Lewis K.O. An analytical method for the quantitation of mannitol and disaccharides in serum: a potentially useful technique in measuring small intestinal permeability *in vivo*. *Clinical Chimica Acta*, 1997, 263 (2),197–205.

1.4 EMERGING PREBIOTICS

Dimitrios K. Tzimorotas and Robert A. Rastall
Department of Food Biosciences
School of Chemistry
Food Bioscience and Pharmacy
University of Reading
Reading
UK

1.4.1   Introduction

Japan has historically been the worldwide leader in production of food and drink products containing functional oligosaccharides (1). They are approved under the Japanese Food for Specified Health Use (FOSHU) regulations as ingredients with beneficial health effects. In the future they may well find their way into the European food market as they offer useful properties and potential benefits as functional ingredients. For this reason we have classed them as "emerging prebiotics".

Most of the oligosaccharides covered in this chapter are not supported by good quality data in human trials. Many of the early trials on these materials were uncontrolled and only used small numbers of individual volunteers. Further, the microbiology in these trials was carried out using culture-based techniques using selective media. This approach is not adequate to characterise the intestinal microbiota, and modern approaches use more robust and representative DNA-based approaches.

This chapter will cover the description, general properties and research studies focused on alternation of intestinal microbiota and health benefits, of the following emerging prebiotic oligosaccharides: isomalto-oligosaccharides (IMO), soya-bean oligosaccharides (SOS), xylo-oligosaccharides (XOS), isomalt, lactosucrose (LS) and other potential prebiotic oligosaccharides.

1.4.2 Emerging prebiotics tested in human trials

1.4.2.1 Isomalto-oligosaccharides

1.4.2.1.1 Manufacture and chemistry

Isomalto-oligosaccharides (IMO) are produced commercially from starch by a two-stage enzymic process (2). First starch is liquefied by α-amylase, then β-amylase hydrolyses starch to maltooligosaccharides, and α-glucosidase catalyses the transglycosylation converting the α1-4 linkages to α1-6. The IMO produced contain mainly α1-6 linkages with a degree of polymerisation (DP) from 2 to 6, and oligosaccharides with α1-4 and α1-6 bonds, such as panose (Glu α1-6 Glu α1-4 Glu).

TABLE 1.4.I
Isomalto-oligosaccharide composition

Component	Degree of Polymerisation	% Dry weight
Glucose	1	1.8
Maltose	2	5.1
Isomaltose	2	48.8
Other disaccharides	2	3.7
Panose	3	6.9
Isomaltotriose	3	16.9
Other trisaccharides	3	1.6
Unspecified tetrasaccharides	4	15.2

IMO can also be manufactured from dextrans (3, 4), although this approach has not yet been commercialised. The dextran route offers potential for control of the molecular weight distribution of the products. This may have advantages in tailoring the physicochemical properties of the product to specific food applications. Dextran-derived IMO show selective fermentation using *in vitro* model systems (5).

Another interesting method of manufacture of IMO uses sucrose and a combination of dextran sucrase and dextranase. The dextran sucrase builds up dextran chains from sucrose, which is then hydrolysed by the dextranase. Goulas *et al* (6) investigated the use of these enzymes in a membrane reactor and found that the product spectrum could be manipulated by controlling reactor operating parameters. The fermentation selectivity of these products has not yet been investigated.

1.4.2.1.2 Fermentation properties

IMO are partially metabolised in the human small intestine (7). In rats they are slowly digested in the jejunum, with the higher DP fractions being less digestible (8). There may, therefore, be an advantage in considering dextrans as sources of

IMO, as the production of higher molecular weight IMO is possible using this route (3, 4). The undigested parts reach the colon where they are metabolised.

According to Kohmoto *et al.* (9), isomaltose, isomaltotriose, panose and the commercial product, Isomalto-9000, are metabolised by different bifidobacteria (*Bifidobacterium longum, Bifidobacterium breve, Bifidobacterium adolescentis* and *Bifidobacterium infantis) Bacteroides, Enterococcus faecalis* and *Clostridium rhamnosum*, but not by other colonic bacteria tested in pure cultures.

In vitro studies using pH-controlled faecal batch cultures showed that IMO are selectively fermented by bifidobacteria (10, 11). Fluorescent *in situ* hybridisation (FISH) was used to enumerate changes in bacterial populations.

Kaneko *et al.* (12) studied the effect of DP *in vivo*, of different IMO products, on their fermentation properties. However only bifidobacteria and total bacterial counts were measured, which does not enable conclusions on their prebiotic effect to be drawn. In another study, Kohmoto *et al.* (13) investigated the dose response of IMO. The minimum effective dose was 8-10 g, determined by a significant increase shown in bifidobacteria using culture-based technique.

As far as health benefits are concerned, IMO supplementation (10g), into ordinary low fibre diets, improved bowel movement and stool output, relieving constipation in seven elderly men (14). Finally, according to Wang *et al.* (15) 20 haemodialysis patients, fed with 30g IMO for 4 weeks, experienced relief from constipation, and their total cholesterol and triglycerides were significantly lowered.

Given the potential shown by IMO *in vitro*, more human trials on these materials would seem to be justified.

1.4.2.2 Soybean oligosaccharides

1.4.2.2.1 Manufacture and chemistry

Soybean oligosaccharides (SOS) are α-galactosyl sucrose derivatives (2). SOS are extracted directly from soybean whey. The main oligosaccharides contained are raffinose (Gal α1-6 Glc α1-2β Fru) and stachyose (Gal α1-6 Gal α1-6 Glc α1-2β Fru). The commercial product from Calpis Food Industry Co. is a concentrated syrup, containing 75% (w/V) solids, of which 35% are SOS (16).

1.4.2.2.2 Fermentation properties

Raffinose and stachyose are resistant to digestion and reach the colon after feeding to humans (17). Fermentation properties have been studied using either individual components, such as raffinose, stachyose, or a mixture of oligosaccharides.

In pure culture, raffinose was metabolised by bifidobacteria and other enteric bacteria, but not by *Lactobacillus acidophilus, Streptococcus faecalis* and *Escherichia coli* (18). In addition, Jaskari *et al.* (19) reported that raffinose was metabolised effectively by *B. infantis, B. bifidum, B. longum, Bact.*

thetaiotaomicron and *Bact. Fragilis*, but poorly by *E. coli* and not at all by *Clostridium difficile*.

In vitro, SOS have been reported to increase faecal bifidobacterial populations (17, 10). However in the study contacted by Rycroft *et al.* (10), an increase in bacteroides was also noticed.

Human studies showing the bifidogenic effect of SOS are summarised in Table 1.4.II. These are, however, non-controlled human feeding studies. Hara *et al.* (20) reported that consumption of 1-2g per day SOS resulted in an increase of faecal bifidobacteria in individuals with initially low numbers of these bacteria.

TABLE 1.4.II
Human studies showing the bifidogenic effect of SOS

Dose/Day	Duration	Subjects	Other effects	Reference
15g	4 weeks	7		21
10g	3 weeks	6	↓ of clostridia	17
0.6g	3 weeks	7	↓ of genotoxic enzymes	22
3g	30 days	8	↑ of propionate, butyrate	23

The above studies have shown the bifidogenic effect of SOS, however there is little information related to selectivity. SOS can also be fermented by other bacteria, sometimes leading to production of intestinal gas, particularly when large doses are used.

As far as health benefits are concerned, raffinose consumption has been related to allergy prevention. According to Nagura *et al.*, (24), a diet supplemented with 50g raffinose in mice, may result in suppression of serum immunoglobulin E response. More specifically, suppressed Th2-type immune response against oral antigen in the lymphoid organs located in or near the intestine.

1.4.2.3 Xylo-oligosaccharides

1.4.2.3.1 Manufacture and chemistry

Xylo-oligosaccharides (XOS) are low molecular weight reducing oligosaccharides, and they are used in soft drinks due to their acid stability. Commercial XOS products are powders and syrup (Suntory Ltd) (16).

Xylan from corn cobs is used commercially for the production of XOS, but xylan from oats or wheat arabinoxylan could also be used. Xylan is hydrolysed by endo-1,4-β-xylanase to produce xylobiose with small amounts of other oligosaccharides. Xylose and other high molecular weight carbohydrates are removed by ultrafiltration and reverse osmosis (2). The end product consists of linear β1-4 linked XOS and oligosaccharides with branched arabinose residues.

1.4.2.3.2 Fermentation properties

Xylan is classed as a dietary fibre and reaches the colon intact. There are several *in vitro* studies on bacterial metabolism of XOS. According to Okazaki *et al.* (25), XOS were metabolised by a range of bifidobacteria and lactobacilli in pure cultures, but not by some other bacteria such as *Bacteroides* and *Clostridium butyricum*. In another pure culture study (19), XOS from oat spelt xylan, were metabolised not only by bifidobacteria, but also by *Bacteroides*, *C. difficile* and *E. coli*, but not by lactobacilli. Finally, the bifidogenic effect of XOS was shown in several other *in vitro* studies (10, 26).

Recently Hughes *et al.* (27) studied the fermentation selectivity of arabinoxylan fractions of varying molecular weight *in vitro*. They used wheat arabinoxylan fractions with average molecular weights of 66, 278 and 354. They found that the selectivity for health-positive bacteria was size-dependent, with the low molecular weight fraction having highest selectivity for bifidobacteria.

In a study with rats (28), a larger increase in bifidobacteria was induced by XOS compared with the same dose of FOS. In a human feeding trial (25), nine adults were fed 5 g per day XOS for three weeks. A range of intestinal bacteria were examined, including pathogens. Only bifidobacteria and *Megasphaera* spp. numbers were increased.

There has been interest recently in oligosaccharides manufactured from wheat arabinoxylans (29). This group fractionated a wheat pentosan concentrate to isolate arabinoxylo-oligosaccharides with an average degree of polymerisation of 3, 5, 12, 15 and 61. These fractions had arabinose:xylose ratios of 0.26, 0.27, 0.69, 0.27 and 0.58, respectively. When fed to rats these fractions were bifidogenic, particularly those with lower arabinose substitution. DP Fractions with a higher DP reduced branched chain fatty acids and elevated acetate, propionate and butyrate. This pattern is suggestive of a shift away from protein metabolism to carbohydrate metabolism.

Based on *in vitro* studies, XOS would seem to be a very promising candidate for a prebiotic, with low active dose and good stability. More well-designed human studies are essential in order to evaluate the prebiotic activity and status of XOS.

1.4.2.4 Isomalt

1.4.2.4.1 Manufacture and chemistry

Isomalt is a trademark given to a mixture of two components: 1-*O*- α-D-glucopyranosyl-D-mannitol and 1-*O*- β -D-glucopyranosyl-D-glucitol. It is made by a two-step process. Firstly sucrose is converted to isomaltulose (6-*O*-α-D-glucopyranosyl-D-fructofuranose, or palatinose) using bacterial cells. The palatinose is then catalytically reduced to the polyol mixture. Isomalt is manufactured commercially by Südzucker in Germany.

1.4.2.4.2 Fermentation properties

Isomalt was well tolerated in human volunteer trials at 30 g per day (30), and did not result in any adverse gastrointestinal effects. Gostner *et al.* (31) carried out a human volunteer trial to investigate the potential prebiotic properties of Isomalt. Nineteen healthy volunteers consumed 30 g of Isomalt or sucrose daily in a double-blind, crossover study. A comprehensive panel of fluorescent *in situ* hybridisation (FISH) probes were used to characterise the microbiota of the volunteers. Significant increases in bifidobacteria and decreases in bacteroides were seen. Faecal β-glucosidase activity decreased, but not β-glucuronidase, sulphatase, nitroreductase or urease.

1.4.2.5 Lactosucrose

1.4.2.5.1 Manufacture and chemistry

Lactosucrose (LS) is a non reducing trisaccharide (Gal β1-4 Glc α1-2β Fru), produced by a mixture of sucrose and lactose. β-fructofuranosidase is used so that the fructosyl residue, from sucrose, to be transferred to the glucose moiety of lactose (16). Different powders and syrups are available, with commercial LS products being made by Ensuiko Sugar Refining Co. and Hayashibara Shoji Inc. (16).

1.4.2.5.2 Fermentation properties

LS reported to be fermented well *in vitro* by bifidobacteria (18). In a human trial conducted by Ohkusa *et al.* (32), the diet of healthy adults was supplemented with 3g per day LS, and faecal bacteria were enumerated on agars. Bifidobacteria levels significantly increased, and bacteroides significantly decreased in comparison with the initial levels of the same bacteria. However, short chain fatty acid levels remained unchanged.

In another human feeding trial, constipated patients were given LS at a dose of 0.32 g per kg of their body weight per day (33). Faecal bacteria were enumerated on agars. A significant increase in bifidobacteria levels, and a large and significant decrease in clostridia were found. Short chain fatty acid levels remained unchanged except acetate and butyrate, which increased.

In contrast, a study involving LS supplementation at 8.5 g per day on patients with chronic inflammatory bowel disease did not result in any increase in bifidobacterial levels (34).

Finally, Hara *et al.* (20), reported a decrease in faecal clostridia and bacteroides numbers followed by LS consumption, as well as a reduction in toxic enzyme activities.

Several potential health benefits as a result of the administration of LS, have been reported, such as enhancement of intestinal calcium absorption in healthy young women with lower than recommended calcium intakes (35), increased intestinal calcium absorption in growing rats (36), and prevention of IgE-mediated allergic diseases in mice (37).

1.4.3 Potential prebiotic carbohydrates not yet tested in humans

A range of other oligosaccharides have been reported as potential prebiotics based on data from *in vitro* model systems. No human data have been obtained yet for these compounds.

1.4.3.1 Gluco-oligosaccharides

1.4.3.1.1 Manufacture and chemistry

Gluco-oligosaccharides can be produced using enzyme glucosyl transfer reactions (2, 38). Dextransucrase acts on sucrose in the presence of maltose as an acceptor carbohydrate. They are enriched in $\alpha 1 \rightarrow 2$-linkages. They can also be made through fermentation with *Leuconostoc mesenteroides*.

1.4.3.1.2 Fermentation properties

According to Djouzi *et al.* (39), *B. breve*, *B. longum*, *Bifidobacterium pseudocatenulatum*, *Bacteroides* spp., and *Clostridium* spp. utilised glucooligosaccharides, but not *B. bifidum* and lactobacilli. These researchers also fed gluco-oligosaccharides to gnotobiotic rats inoculated with human faecal bacteria. A limited number of microbial groups were counted, and the microbiology was performed using selective media. No changes were seen in microbial populations. Chung and Day (40) produced branched chain oligomers, using *Leuconostoc mesenteroides* B-742, which were metabolised by bifidobacteria and lactobacilli in pure cultures, but not by *Salmonella* spp. or *E. coli*.

1.4.3.2 Pectic oligosaccharides

Pectic oligosaccharides have recently attracted a lot of interest for their varied biological potential. They can be made by enzymic hydrolysis of pectins (41), or by flash extraction of citrus processing waste (42). They have been evaluated for their prebiotic potential using *in vitro* batch culture systems (43, 44). Both enzymically derived pectin (from citrus and apple) and flash extracted pectic oligosaccharides have shown promise as selective substrates for bifidobacteria *in vitro* (43, 44). Further, they have the ability to inhibit the action of *E. coli* verocytotoxins against human colonic cells in culture (45, 46), and to inhibit the adhesion of certain pathogens to the same cells in culture (46).

71

1.4.3.3 Gentio-oligosaccharides

Gentio-oligosaccharides are $\alpha1\rightarrow6$-linked gluco-oligosaccharides with a DP of between two and six (16). Rycroft *et al.* (47) studied the selectivity of fermentation using simple *in vitro* batch cultures with a human faecal inoculum. They proved to be selective towards bifidobacteria and to give rise to elevated short chain fatty acid levels. Different fractions of gentio-oligosaccharides were studied for their selectivity *in vitro* (48) and a preference for oligosaccharides with a DP of three was observed.

The prebiotic potential of the above and of many other candidate prebiotic saccharides should be investigated more thoroughly. *In vitro* and well designed *in vivo* studies are necessary in order to be able to evaluate any prebiotic effect.

1.4.4 Conclusions and future developments

So far there is a promising evidence for prebiotic activity of a range of oligosaccharides mentioned above. The available literature covers different *in vitro* and *in vivo* studies. Pure cultures studies are focused mainly on *Bifidobacterium* spp. and *Lactobacillus* spp. Prebiotic oligosaccharides should be fermented by bifidobacteria and lactobacilli more than the other types of bacteria. However, in many cases a wide range of bifidobacteria and lactobacilli were tested, but only a few strains from the other potential pathogenic species. Thus the evaluation of the selectivity of the oligosaccharide is not possible. Moreover, pure cultures can only offer an indication about the faith of oligosaccharides in the colon. Mixed culture conditions in the colon affect the fermentation behaviour.

Unfortunately there are only a few well designed human studies reported in the literature. Many of them are uncontrolled feeding studies, often using only a few individuals, and carried out for short periods of time. In addition, selective agars were used for bacterial enumeration and assessment of selectivity. There is a need for well designed human trials that will evaluate more efficiently the prebiotic potential of the tested substrates, and clinically demonstrate benefits to health.

In the future, with the use of new molecular microbiological techniques, more non-cultivable bacteria will be enumerated and taken into consideration for the evaluation of the colonic microbiota changes induced by the tested carbohydrate. More detailed information about species rather than genera will also be gained.

In this way our knowledge of the gut flora diversity will be increased and conclusions related to structure-function relationships may be drawn. So it may become possible to design selectivity or use different carbohydrates in order to achieve several health benefits.

Moreover, nowadays bifidobacteria and lactobacilli are the main target groups of prebiotic carbohydrates, but the list of beneficial bacteria may be increased. It should also be highlighted that alteration of colonic microbiota is only a part of prebiotic action. Proliferation of other groups of bacteria that may offer health benefits, such as butyrate production and immunomodulation, is also desirable.

Finally, the evaluation of the prebiotic potential of a carbohydrate should be accompanied by an investigation of the potential of it being incorporated as a food ingredient. Physicochemical and organoleptic properties of carbohydrates should not be neglected. They are affected by monosaccharide composition, structures and molecular weight.

To sum up, it can be said that research and efforts are still needed so that emerging prebiotics establish their place in food industry market and gain consumers' confidence.

1.4.5   References

1. Nakakuki T. Present status and future of functional oligosaccharide development in Japan. *Pure and Applied Chemistry*, 2002, 74, 1245-51.

2. Crittenden R.G.,Playne M.J. Production, properties and applications of food-grade oligosaccharides. *Trends in Food Science and Technology*, 1996, 7, 353-61.

3. Mountzouris K.C., Gilmour S.G., Grandison A., Rastall R.A. Modelling of oligodextran production in an ultrafiltration stirred cell membrane reactor. *Enzyme and Microbial Technology*, 1998, 24, 75-85.

4. Mountzouris K.C., Gilmour S.G., Rastall R.A. Continuous production of oligodextrans via controlled hydrolysis of dextran in an enzyme membrane reactor. *Journal of Food Science*, 2002, 67 (5), 1767-71.

5. Olano-Martin E., Mountzouris K.C., Gibson G.R., Rastall R.A. *In vitro* fermentability of dextran, oligodextran and maltodextrin by human gut bacteria. *British Journal of Nutrition*, 2000, 83, 247-55.

6. Goulas A.K., Fisher D.A., Grimble G.K., Grandison A.S., Rastall R.A. Synthesis of isomaltooligosaccharides and oligodextrans by the combined use of dextransucrase and dextranase. *Enzyme and Microbial Technology*, 2004, 35, 327-38.

7. Oku T., Nakamura S. Comparison of digestibility and breath hydrogen gas excretion of fructo-oligosaccharides, galactosyl-sucrose and isomalto-oligosaccharide in healthy human subjects. *European Journal of Clinical Nutrition*, 2003, 57, 1150-6.

8. Kaneko T., Yokoyama A., Suzuki M. Digestibility characteristics of isomaltooligosaccharides in comparison with several saccharides using the rat jejunum loop method. *Bioscience, Biotechnology and Biochemistry*, 1995, 59 (7), 1190-4.

9. Kohmoto T., Fukui F., Takaka H., Machida Y., Arai M., Mitsuoka T. Effect of isomalto-oligosaccharides on human fecal flora. *Bifidobacteria Microflora*, 1988, 7, 61-9.

10. Rycroft C.E., Jones M.R., Gibson G.R., Rastall R.A. A comparative *in vitro* evaluation of the fermentation properties of prebiotic oligosaccharides. *Journal of Applied Microbiology*, 2001, 91, 878-87.

11. Palframan R.J., Gibson G.R., Rastall R.A. Effect of pH and dose on the growth of gut bacteria on prebiotic carbohydrates *in vitro*. *Anaerobe*, 2002, 8, 287-92.

12. Kaneko T., Kohmoto T., Kikuchi H., Shiota M., Iino H., Mitsuoka T. Effects of isomaltooligosaccharides with different degrees of polymerisation on human faecal bifidobacteria. *Bioscience, Biotechnology and Biochemistry*, 1994, 58, 2288-90.

13. Kohmoto T., Fukui F., Takaku H., Mitsuoka T. Dose response test of isomaltooligosaccharides for increasing faecal bifidobacteria. *Agricultural and Biological Chemistry*, 1991, 55, 2157-9.

14. Chen H.L., Lu Y.H., Lin J.J., Ko L.Y. Effects of isomalto-oligosaccharides on bowel functions and indicators of nutritional status in constipated elderly men. *Journal of the American College of Nutrition*, 2001, 20 (1), 44-9.

15. Wang H.F., Lim P.S., Kao M.D., Chan E.C., Lin L.C., Wang N.P. Use of isomalto-oligosaccharide in the treatment of lipid profiles and constipation in hemodialysis patients. *Journal of Renal Nutrition*, 2001, 11, 73-9.

16. Playne M.J., Crittenden R. Commercially available oligosaccharides. *Bulletin of the International Dairy Federation*, 1996, 313, 10-22.

17. Hayakawa K., Mizutani J., Wada K., Masai T., Yoshihara I., Mitsuoka T. Effects of soybean oligosaccharides on human faecal flora. *Microbial Ecology in Health and Disease*, 1990, 3, 293-303.

18. Minami Y., Yazawa K., Tamura Z., Tanaka T., Yamamoto T. Selectivity of utilisation of galactosyl-oligosaccharides by bifidobacteria. *Chemical and Pharmaceutical Bulletin*, 1983, 31, 1688-91.

19. Jaskari J., Kontula P., Siitonen A., Jousimies-Somer H., Mattila-Sandholm T., Poutanen K. Oat β-glucan and xylan hydrolysates as selective substrates for *Bifidobacterium* and *Lactobacillus* strains. *Applied Microbiology and Biotechnology*, 1998, 49, 175-81.

20. Hara T., Ikeda N., Hatsumi K., Watabe J., Lino H.,Mitsuoka T. Effects of small amount ingestion of soybean oligosaccharides on bowel habits and faecal flora of volunteers. *Japanese Journal of Nutrtion*, 1997, 55, 79-84.

21. Benno Y., Endo K., Shiragami N., Sayama K., Mitsuoka T. Effects of raffinose intake on human faecal microflora. *Bifidobacteria Microflora*, 1987, 6, 59-63.

22. Wada K., Watabe J., Mizutani J., Tomoda M., Suzuki H., Saitoh Y. Effects of soybean oligosaccharides in a beverage on human fecal flora and metabolites. *Journal of Agricultural Chemical Society of Japan.* 1992, 66, 127-135.

23. Bang M.H., Chio O.S.,Kim W.K. Soyoligosaccharide increases faecal bifidobacteria counts, short chain fatty acids and faecal lipid concentrations in young Korean women. *Journal of Medicinal Food*, 2007, 10 (2), 366-70.

24. Nagura T., Hachimura S., Hashiguchi M., Ueda Y., Kanno T., Kikuchi H., Sayama K., Kaminogawa S. Suppressive effect of dietary raffinose on T-helper 2 cell-mediated immunity. *British Journal of Nutrition*, 2002, 88, 421-7.

25. Okazaki M., Fujikawa S., Matsumoto N. Effects of xylooligosaccharide on growth of bifidobacteria. *Journal of the Japanese Society of Nutrition and Food Science*, 1990, 43, 395-401.

26. Van Laere K.M., Hartemink R., Bosveld M., Schols H.A., Voragen A.G. Fermentation of plant cell wall derived polysaccharides and their corresponding oligosaccharides by intestinal bacteria. *Journal of Agricultural and Food Chemistry*, 2000, 48, 1644-52.

27. Hughes S.A., Shewry P.R., Li L., Gibson G.R., Sanz M.L., Rastall R.A. *In vitro* fermentation by human faecal microflora of wheat arabinoxylans. *Journal of Agricultural and Food Chemistry*, 2007, 55, 4589-95.

28. Campbell J.M., Fahey G.C.,Wolf B.W. Selected indigestible oligosaccharides affect large bowel mass, cecal and faecal short-chain fatty acids, pH and microflora in rats. *Journal of Nutrition*, 1997, 127, 130-6.

29. Van Craeveld V., Swennen K., Dornez E., Van de Wiele T., Marzorati M., Verstraete W., Delaedt Y., Onagbesan O., Decuypere E., Buyse J., De Ketelaere B., Broekaert W.F., Delcour J.A., Courtin C.M. Structurally different wheat-derived arabinoxylooligosaccharides have different prebiotic and fermentation properties in rats. *Journal of Nutrition*, 2008, 138, 2348-55.

30. Gostner A., Schaffer V., Theis S., Menzel T., Luhrs H., Melcher R., Schauber J., Kudlich T., Dusel G., Dorbath D., Kozianowski G., Scheppach W. Effects of isomalt consumption on gastrointestinal and metabolic parameters in healthy volunteers. *British Journal of Nutrition*, 2005, 94, 575-81.

31. Gostner A., Blaut M., Schaffer V., Kozianowski G., Theis S., Klingeberg M., Dombrowski Y., Martin D., Ehrhardt S., Taras D., Schwiertz A., Kleesen B., Luhrs H., Schauber J., Dorbath D., Menzel T.,Scheppach W. Effect of isomalt consumption on faecal microflora and colonic metabolism in healthy volunteers. *British Journal of Nutrition*, 2006, 95, 40-50.

32. Ohkusa T., Ozaki Y., Sato C., Mikuni K., Ikeda H. Long term ingestion of lactosucrose increases Bifidobacterium sp. in human faecal flora. *Digestion*, 1995, 56, 415-20.

33. Kumemura M., Hashimoto F., Fujii F., Matsuo K., Kimura H., Miyazoe R., Okamatsu H., Inokuchi T., Ito H., Oizumi K., Oku T. Effects of administration of 4G-β-D-galactosylsucrose on faecal microflora, putrefactive products, short chain fatty acids, weight, moisture and pH and the subjective sensation of defecation in the elderly constipation. *Journal of Clinical Nutrition*, 1992, 13, 199-210.

34. Teramoto F., Rokutan K., Kawakami Y., Fujimura Y., Uchida J., Oku T., Oka M., Yoneyama M. Effect of 4G-beta-D-galactosylsucrose (lactosucrose) on faecal microflora in patients with chronic inflammatory bowel disease. *Journal of Gastroenterology*, 1996, 31, 33-9.

35. Teramoto F., Rokutan K., Sugano Y., Oku K., Kishino E., Fujita K., Kishi K., Fukunaga M., Morita T. Long-term administration of 4G-β-D-galactosylsucrose (lactosucrose) enhances intestinal calcium absorption in young women: a randomized, placebo-controlled 96-wk study. *Journal of Nutritional Science and Vitaminology*, 2006, 52, 337–46.

36. Kishino E., Norii M., Fujita K., Hara K., Teramoto F., Fukunaga M. Enhancement by lactosucrose of the calcium absorption from the intestine in growing rats. *Bioscience, Biotechnology and Biochemistry*, 2006, 70 (6), 1485-8.

37. Taniguchi Y., Mizote A., Kohno K., Iwaki K., Oku K., Chaen H., Fukuda S. Effects of dietary lactosucrose (4G-β-D-galactosylsucrose) on the IgE response in mice. *Bioscience, Biotechnology and Biochemistry*, 2007, 71 (11), 2766-73.

38. Dols M., Monchois V., Remaud-Siméon M., Willemot R-M.,Monsan P. The production of α1→2-terminated glucooligosaccharides, in *Methods in Biotechnology, Carbohydrate Biotechnology Protocols*. Ed. Bucke C., Humana Press, 1999, 10, 129-39.

39. Djouzi Z., Andrieux C., Pelenc V., Somarriba S., Popot F., Paul F., Monsan P., Szylit O. Degradation and fermentation of alpha-gluco-oligosaccharides by bacterial strains from human colon: *in vitro* and *in vivo* studies in gnotobiotic rats. *Journal of Applied Bacteriology*, 1995, 79, 117-27.

40. Chung C.H., Day D.F. Glucooligosaccharides from *Leuconostoc mesenderoides* B-742 (ATCC 13146): A potential prebiotic. *Journal of Industrial Microbiology and Biotechnology*, 2002, 29 (4), 196-9.

41. Olano-Martin E., Mountzouris K.C., Gibson G.R., Rastall R.A. Continuous production of oligosaccharides from pectin in an enzyme membrane reactor. *Journal of Food Science*, 2001, 66 (7), 966-71.

42. Fishman M.L., Chau H.K., Hoagland P., Ayyad K. Characterization of pectin, flash-extracted from orange albedo by microwave heating, under pressure. *Carbohydrate Research*, 2003, 323 (1-4), 126-38.

43. Olano-Martin E., Gibson G.R., Rastall R.A. Comparison of the *in vitro* bifidogenic properties of pectins and pectic oligosaccharides. *Journal of Applied Microbiology*, 2002, 93, 505-11.

44. Manderson K., Pinart M., Tuohy K.M., Grace W.E., Hotchkiss A.T., Widmer W., Yadhav M.P., Gibson G.R., Rastall R.A. *In vitro* determination of the prebiotic properties of oligosaccharides derived from an orange juice manufacture byproduct stream. *Applied and Environmental Microbiology*, 2005, 71 (12), 8383-9.

45. Olano-Martin E., Williams M.R., Gibson G.R., Rastall R.A. Pectins and pectic-oligosaccharides inhibit *Escherichia coli* O157:H7 Shiga toxin as directed towards the human colonic cell line HT29. *FEMS Microbiology Letters*, 2003, 218 (1), 101-5.

46. Rhoades J., Manderson K., Wells A., Hotchkiss Jr. A.T., Gibson G.R., Formentin K., Beer M, Rastall R.A. Oligosaccharide-mediated inhibition of the adhesion of pathogenic Escherichia coli strains to human gut epithelial cells in vitro. *Journal of Food Protection*, In press.

47. Rycroft C.E., Jones M.R., Gibson G.R., Rastall R.A. Fermentation properties of gentio-oligosaccharides. *Letters in Applied Microbiology*, 2001, 32, 156-61.

48. Sanz M.L., Côté G.L., Gibson G.R., Rastall R.A. Selective fermentation of gentiobiose derived oligosaccharides by human gut bacteria: Influence of molecular weight. *FEMS Microbiology Ecology*, 2006, 56, 383-8.

2. PROBIOTICS

2.1 BIFIDOBACTERIA

Margaret O'Connell, MSc.
Chr. Hansen Ltd.
2 Tealgate
Hungerford
RG17 0YT
UK

2.1.1   Introduction

Probiotics have been defined as "live microorganisms that when ingested in adequate amounts, confer a health benefit on the host" (1), which indicates the criteria that bacteria must meet in order to be considered as a probiotic. The key genera of bacteria implicated as probiotics are *Lactobacillus* and *Bifidobacterium*. This chapter will focus on the *Bifidobacteria* and their potential benefits to consumers.

2.1.2   Description

2.1.2.1   Taxonomy of bifidobacteria

Bifidobacteria were first isolated from the faeces of a healthy child by Tissier circa. 1900, who classified this new isolate as *Bacillus bifidus* (2, 3). The organism was later reclassified as *Lactobacillus bifidus*. The genus *Bifidobacterium* was proposed by Orla-Jensen in 1924 (4). Over the passing years, bifidobacteria have been classified as *Bacillus, Lactobacillus, Nocardia,* corynebacteria and *Bacteroides*, but mainly as lactobacilli. It was not until 1974 that these organisms were reclassified as bifidobacteria, due to application of modern taxonomic tools which highlighted their unique metabolic pathways, DNA hybridisation and GC content (5).

Bifidobacteria are now classified as: phylum *Actinobacteria*, class *Actinobacteria*, sub-class *Actinobacteridae*, order *Bifidobacteriales*, family *Bifidobacteriaceae* (6), genus *Bifidobacterium*. Currently, upwards of 30 species are reported across the literature (7). Species included among the bifidobacteria are:*B. adolescentis, B. angulatum, B. animalis* subsp. *lactis, B. animalis* subsp. *animalis, B. bifidum, B. catenulatium, B. dentium, B. gallicium, B. longum* biotype *infantis, B. longum* biotype longum, *B. pseudocatenulatum* and *B. scardovi. B.*

animalis subsp. *animalis* and *B. animalis* subsp. *lactis* represent two new subspecies resulting from the reclassification of *B. animalis* (8).

Bifidobacteria have been isolated from many sources such as sewage, human and animal faeces, the rumen of cattle, dental caries and also from honey bees (9). Their most significant commercial and scientific application is as dietary probiotics.

2.1.2.2 Physiology of bifidobacteria

Bifidobacteria are Gram-positive, polymorphic rods. They characteristically display 'Y' or 'V' shaped cell morphology, hence the term 'bifid', but can also be club or spatulate shaped. They are non-sporeforming, non-motile and catalase negative. They are also obligate anaerobes, although some species have been shown to tolerate low levels of oxygen, for example, Meile *et al.* (10) characterised *B. animalis* subsp. *lactis* that tolerated 10% oxygen.

One key characteristic of the bifidobacteria is their hexose metabolic pathway, known as the 'bifidus shunt'. Essentially, this is where lactose, or other suitable fermentable carbohydrates are converted to lactate and acetate, which is produced in a 2:3 ratio respectively (11). The key enzyme involved in this pathway is fructose-6-phosphoketolase and it appears to be specific to the bifidobacteria. It is often used as an identifying factor (12). This pathway differs from the LAB (Lactic Acid Bacteria), which use the glucose-6-phosphate shunt. Additionally, bifidobacteria are able to catabolise a wide range of mono-, di- and oligo-saccharides, which gives them a competitive advantage in the digestive tract. They do not produce gas (13).

Bifidobacteria also have the ability to synthesise various water-soluble vitamins. *B. lactis*, *Bifidobacterium infantis*, *Bifidobacterium breve* and *B. animalis* have been shown to increase the level of folate present in reconstituted skimmed milk fermentation; however, fermentation times ranged from 14 to 40 hours (14). It was reported by Deguchi *et al.* that bifidobacteria can excrete a range of water soluble vitamins, such as nicotinic acid, folate and thiamine (15) However, the ability, type and quantity of vitamin produced vary between strains.

Optimal growth temperatures for the majority of bifidobacterial strains have been reported as between 36 and 38 °C for strains of human origin, while strains sourced from animals tend to have a higher optimum of 41-43 °C (16).

2.1.2.3 Bifidobacteria and prebiotics

A prebiotic is typically an oligosaccharide that serves as a food source for probiotic organisms within the gastro-intestinal tract. They have been defined thus: *'a prebiotic is a selectively fermented ingredient that allows specific changes, both in the consumption and/or activity in the gastrointestinal microflora, that confer benefits upon host well-being and health'* (17). Vernazza *et al.* (19) studied the ability of 5 Bifidobacteria (*B. adolescentis* DSM20083, *B.*

longum DSM20219, *B. longum* 46, *B. infantis* DSM20088 and *B. animalis* subsp. BB12) to grow on 12 different carbohydrate substrates, which included 7 commercially available oligosaccharides, a prebiotic polysaccharide, two putatively prebiotic molecules (lactitol and polydextrose) and 2 non-prebiotic controls, a simple sugar and a polysaccharide (glucose and maltodextrin respectively). Galacto-oligosaccharides (GOS) and isomalto-oligosaccharides (IMO) were well utilised by all of the bifidobacterium tested. None of the bifidobacterium grew on lactitol, whereas *B. animalis* subsp. *lactic* BB12, *B. adolescentis* DSM20083 and *B. longum* DSM20219 were able to grow on high and low molecular weight inulin. The bifidobacterium strains tested exhibited a wide variety of growth rates on the 12 carbohydrates, ranging from 0 to 0.25/h, as determined by optical density.

Essentially, prebiotics seek to improve gut flora composition by acting as a food base for the desirable bacteria already present in the gut, and are often regarded as a practical and convenient way to influence to the gut flora. However, if the desired probiotic bacteria are absent from the gut, then the prebiotic is unlikely to be effective (20).

Therefore, combining prebiotics with probiotic bacteria (synbiotics) is perhaps a way forward for maximum benefit to the consumer. Various carbohydrates have been studied for their probiotic effects, most extensively, fructo-, galacto-, xylo-oligosaccharides and inulin (21-23).

2.1.3 Physiological properties

2.1.3.1 Bifidobacteria and the human gastrointestinal tract

The human gastrointestinal tract contains a large microflora with an estimated number of microorganisms between 10^{13} - 10^{14} (24, 25), most of which are anaerobic and situated in the large intestine. It is a very complex environment and while some bacterial species live there permanently, others are introduced via food and just pass through the digestive tract. In healthy adults, bifidobacteria appear to constitute 3-6% of the native microflora (26, 27). The proportion represented by bifidobacterial species varies between individuals, but also depends on lifestyle factors, such as diet and exercise, and also on age/stage of life.

Researchers have characterised the composition of the gut microflora and the impact of aging on the density and diversity of bacteria found in the gut (26). Newborn infants tend to have an initial microflora constituted largely of *enterobacteriaceae* and *streptococci*; however, after weaning changes occur in the microflora (28). After 1 week of life, fully breast fed infants have been shown to have gut flora composed almost exclusively of *Bifidobacterium* spp., whilst formula fed infants show a more varied flora (29, 30). However, recent work has shown that another species, *Ruminococcus*, may be as prevalent as *Bifidobacterium* spp. (31, 32). The infant formula market has followed this finding by developing products that contain probiotics and/or prebiotics.

As we age and develop an adult gut flora, various bacteria manifest within the gut. The more prominent species found are: *Bacteroides, Enterococci, Escherichia coli, Lactobacilli* and *Bifidobacterium* (33, 34). Bifidobacterial numbers in the gut often decrease substantially with age and/or illness. Hopkins *et al.* studied the composition of the gut flora of a number of different population groups (26). They found that elderly subjects receiving antibiotic treatment had a reduced gut flora both in terms of total cell count, and species diversity compared to healthy elderly subjects. Poor dentition, decrease in appetite and sense of taste and smell lead to changes in diet that may result in a sub-optimal gut flora (35).

However, part of this research was skewed by the results obtained from one 67-year old male, who was unusually active and health conscious, and displayed a high bifidobacterial cell count, which was more in keeping with the healthy, young adult group instead of the healthy, elderly adult group, thus inferring that diet and lifestyle significantly impact on gut composition, rather than just aging itself (26).

Overall, bifidobacterial cell numbers have a tendency to decline with age, due in part to changes in lifestyle and diet. This decline in facultative anaerobes in the aging gut appears to occur side by side with increases in *clostridia* and other negatively perceived bacteria.

2.1.3.2 *Passage of bifidobacteria through the gastointestinal tract*

The function of the digestive tract is to break down food into a format that can be readily assimilated by the body. Thus any probiotic bacteria entering the digestive tract have a number of challenges to meet prior to reaching the colon. Many bacteria lack the ability to pass unscathed through the digestive process. However, the ability to resist digestion is a key feature of any bacteria that would be considered as.a probiotic (36).

2.1.3.2.1 *Acid and bile tolerance of Bifidobacterium*

The hurdles posed by the gastro-intestinal tract are: (i) the acidity of the stomach, which secretes approximately 3 litres of gastric acid (pH ~2.0) daily and (ii) bile salt exposure in the jejunum, both of which can be lethal to bacteria.

Vernazza *et al.* (18) studied the acid and bile tolerance, and fermentable carbohydrate preference of selected strains of Bifidobacterium. The strains were exposed to pH 2.0, 3.0 and 4.0 in a glycine-HCL buffer for periods up to 20 minutes. Most of the Bifidobacterium strains tested were poorly resistant to strongly acidic conditions (pH 2.0), as *B. adolescentis* DSM20083, *B. longum* DSM20219 and *B. longum* 46 failed to survive at the pHs tested. Therefore, encapsulation of these particular bacterial strains would aid their application as probiotics. *B. infantis* DSM20088 was able to survive pHs greater than 4.0. The one acid-tolerant exception was *B. animalis* subsp. *lactis* BB12, which withstood pHs as low as 2.0 for up to 20 minutes.

The bifidobacterium were also tested for their resistance to bile salts, by growing them with +/- 0.5% (w/v) ox-gall at pH 8.0. Both *B. animalis* subsp. *lactis* BB12 and *B. infantis* DSM20088 attained cell counts of >1.0 x 10⁸ cfu/ml in the presence of ox-gall, whilst *B. adolesentis* DSM20083 and *B. longum* DSM20219 and *B. longum* 46 reached cell counts of <4.0 x 10⁴ cfu/ml (18).

However, other researchers have reported inhibition of growth by bile salts. Grill *et al.* (37) showed a strong inhibitory effect across a range of species; *B. longum, B. infantis, B. breve* and *B. animalis*. Perrin *et al.* (38) showed a decreased growth rate of *B. breve, B. longum* and *B. animalis* in the presence of bile salt; however, inclusion of neosugars alleviated some of their bactericidal effect in *B. breve* and *B. longum*.

2.1.3.2.2 Ability of Bifidobacterium to temporarily colonise the gut

Following successful passage through the small intestine, the bacteria now pass into the colon, where they are best adapted to compete. Many bifidobacteria have the ability to bind to colonic cell lines and to mucin (39-42), which is thought to aid their temporary colonisation of the gut. Bifidobacteria have been shown to persist in the gut for up to 6 weeks after consumption has ceased (43).

There are several proposed mechanisms through which probiotics can have an effect. Many probiotics have been shown to inhibit the growth and development of enteropathogenic bacteria, and alter, albeit temporarily, the composition of the gut flora, and modulate the innate immune system. Suggested mechanisms for the effects on host health are: (adapted from de Vrese and Marteau (44))

a. Competitive inhibition of other organisms through competition for fermentable substrates and adhesion sites within the gut

b. Lowering gut pH through production of organic acids, e.g. acetic, lactic and butyric acids

c. Production of bacteriocins

d. Stimulation of the immune system

e. Regulation of gut motility

f. Strengthening gut barrier function

g. Binding and metabolism of toxic substances

h. Release of gut protective metabolites (43)

2.1.4   Specific health benefits associated with *bifidobacteria*

One of the most widely studied and documented aspects of probiotic bacteria is their effect on gastrointestinal health. This is a large area of research, and encompasses many segments of interest, each an area of research in itself, such as effect of probiotics on infantile diarrhoea, gut health in the elderly, gastroenteritis, antipathogenic effects, traveller's diarrhoea, and antibiotic associated diarrhoea (AAD).

2.1.4.1 Effect of bifidobacteria on Enteropathogens

Diarrhoea can arise through various routes, but is largely due to action of undesirable organisms in the gastrointestinal tract. Many of the bacteria involved in gastrointestinal infection have recognisable names, such as *Salmonella* spp., *Listeria* (*monocytogenes*) and *Clostridium difficile*. However, there are many more bacteria that can cause digestive disease, such as *Helicobacter pylori*, which is the causative agent in gastric ulcers, and certain species of *Campylobacter*. The effect of probiotics on preventing the negative action of many of these organisms has been investigated.

Researchers examined the action of bifidobacteria isolated from infant stools on *Listeria monocytogenes* (45). Five of the isolates (out of 34) were found to secrete antibacterial substance(s) which were able to inhibit the *Listeria* strains *in vitro*.

The effect of *B. bifidum* to protect mice against *S. enteritidis* ssp. *typhimurium* in conventional and gnotobiotic mice was studied by Silva *et al.* (46). All mice were challenged with *Salmonella* bacteria, with the test groups also receiving 'Bifidus' milk, which was found to protect both types of mice against the challenge, as shown by survival and histopathological results. However, in order to achieve the protective effect in the gnotobiotic mice, it was necessary to administer the bifidus milk 10 days prior to salmonella challenge. Further investigation of the underlying mechanism of protection of *B. longum* against *Salmonella* was attributed to a reduction in inflammatory response in the mice mediated by the probiotic treatment (47).

Asahara *et al.* (47) observed that *B. breve*-colonised mice were significantly better equipped to resist the pathogenicity of an *E. coli* O157:H7 challenge, compared with the corresponding control group. Gagnon *et al.* (48) found two isolates of *B. bifidum* showed significant potential to inhibit adhesion of *E. coli* O157:H7 to Caco-2 cells.

2.1.4.2 Effect of bifidobacteria on antibiotic-associated diarrhoea

Disturbance of the gastrointestinal flora through antibiotic consumption can lead to proliferation of bacteria such as *Clostridium difficile*, which in turn leads to diarrhoea. This antibiotic-associated diarrhoea (AAD) is a common problem, and is estimated to affect 25-30% of patients receiving antibiotic treatment, with 25% of cases as a result of *C. difficile* proliferation (44).

Plummer *et al.* conducted a trial in which hospitalised patients receiving antibiotics also received probiotic supplementation (49). In the test group, *C. difficile* toxins were found in 46% of the probiotic group compared with 78% in the control group. In another study, Colombel *et al.* (50) found that consumption of yoghurt made with *B. longum* reduced the duration of erythromycin induced diarrhoea.

Additionally, overgrowth of candida species in the gut is often associated with antibiotic consumption. A study on the effect of various probiotics on *Candida albicans* infected, immunodeficient mice showed a marked reduction in the incidence and severity of candidiasis in the mice supplemented with *B. animalis* (51). Another study reported a reduction of candida shed in faeces of chemotherapy patients who consumed a milk drink containing *Bifidobacterium* spp. and *L. acidophilus* (52).

D'Souza *et al.* (53) conducted a meta-analysis across several studies, involving different probiotic agents (some of which included *Bifidobacterium* spp.) and their efficacy in combating AAD. They concluded that probiotics may be useful in prevention of antibiotic associated diarrhoea, but that there was insufficient evidence to support probiotics as a means of treating active diarrhoea.

In 1999, more than 15,000 cases of *C. difficile* were reported in the National Health Service (NHS) in the UK, with a corresponding treatment cost of £60 million (49). Therefore probiotics could be a possibly cost effective aid in treatment.

2.1.4.3 Effect of bifidobacteria on Helicobacter pylori

H. pylori is a common bacterium in the human gastrointestinal tract. It is highly acid-tolerant and is able to survive the low pH environment of the stomach. *H. pylori* is a major cause of chronic gastritis and peptic ulcers (54). There is interest in the potential benefit of probiotics on *H. pylori* in the gut, either as a mean of reducing colonisation or assisting in antibiotic eradication and aiding tolerance of antibiotic treatment.

Researchers showed that *in vitro*, *Bifidobacterium* BB12 had an inhibitory effect against *H. pylori*, also that regular consumption of AB yoghurt (containing *Bifidobacterium* BB12 and *Lb. Acidophilus* LA5) over a 6-week period decreased *H. pylori* activity (55). It was also demonstrated that consumption of AB yoghurt (strains as before) reduced *H. pylori* loads sufficiently to improve efficacy of quadruple therapy in eliminating residual *H. pylori* (after failed triple therapy), compared to the control group who received quadruple therapy only (56).

2.1.4.4 Effect of bifidobacteria on diarrhoea in infants and children

Infants have a gut flora largely based on bifidobacteria, but feeding (breast-fed versus formula-fed) and delivery (conventional versus caesarean section) can have a significant impact on composition, both in species and numbers of bacteria present (57). During the first two years of life, a child will establish a stable gut microflora, consisting predominantly of anaerobic organisms. Due to this immature gut flora and immune system, children are vulnerable to gastrointestinal and other infections, especially in day-care environments, where infections are frequent and pass readily between children (23). Thus, an effective method to prevent gastroenteritis in children is an important goal.

Chouraqui *et al.* studied the effect of *Bifidobacterium* BB12 supplemented, acidified formula on incidence and duration of diarrhoea in 90 infants in residential care environments, in a blinded, randomised trial (58). Results indicated that infants in the *Bifidobacterium* supplemented formula group had a reduced risk of getting diarrhoea, and those patients that did develop it had shorter episodes compared with the control group.

Further research investigated the impact of *B. bifidum* plus *S. thermophilus* supplemented formula in 55 infants with acute diarrhoea (59). The researchers found that consumption of the bacterial supplemented formula resulted in reduced cumulative incidence of diarrhoea during hospitalisation (6.9% in supplemented formula group, compared to 31% of the control group) and, for those who did become ill, a reduction in shedding of rotavirus in the stools post-infection (10.3% for supplemented group, versus 38.5% for the control group).

In a study involving healthy children, Fukushima *et al.* investigated the effect of supplementing follow-on infant formula with *Bifidobacterium lactis* (60). Reduced quantities of putrefactive products such as ammonia and indole were found in faeces, whilst acetate content was increased in the bifidobacterial supplemented group. In all studies the supplemented formulas were well tolerated by subjects.

2.1.4.5 Effect of bifidobacteria on traveller's diarrhoea

Depending on location, it is estimated that the risk of traveller's diarrhoea is approximately 7% in developed countries, while this figure rises to 20-50% risk when travelling to the developing world (61). Consequently, there is interest in the potential use of probiotics to minimise the likelihood of suffering from this malady, or to reduce the severity of an attack. In a double blind, placebo controlled study, Black *et al.* (62) administered capsules containing *Bifidobacterium* BB12, *Lb. acidophilus* LA5, *St. thermophilus* and *Lb. delbrueckii* subsp. *bulgaricus* to a group of 90 Danish tourists on holiday in Egypt. The frequency of traveller's diarrhoea was reduced from 71% in the control group to 43% in the probiotic group.

2.1.4.6 Effect of bifidobacteria on constipation

Constipation is a common problem and is most prevalent in the elderly. As previously mentioned, microflora in the elderly tends to be lower in bifidobacterial content and higher in more putrefactive organisms (63).

Researchers reported on the effect of Cultura (a fermented milk containing *Bifidobacterium* BB12 and *L. acidophilus* LA5) on 23 elderly subjects with a history of chronic constipation (64). Subjects were supplemented with either probiotic fermented or unfermented milk placebo, over 4 periods of 5-6 weeks duration. A significant increase in bowel movement was observed during periods of probiotic milk consumption. Sagen (65) examined the effect of probiotics, in capsule form, on the 24 nursing home patients with various intestinal problems, and found that intestinal function was improved in 56% of the probiotic group compared with 10% in the placebo group. Similarly, another study investigated the effect of consumption of yoghurt with bifidobacteria on elderly subjects who were confined to bed. Those receiving the bifidus yoghurt reported improved stool frequency during yoghurt consumption (66).

2.1.4.7 *Effect of bifidobacteria on immunity and allergy*

It is estimated that approximately 20% of Westernised populations suffer from some form of allergy, and the instance of affected individuals is still increasing. Individuals at most risk carry a genetic predisposition towards allergy, but environmental triggers are also required for allergy development (67).

The early months of life are critical in the development of gut microbiota, during which time, the immature immune system and developing gut microflora interact. However, increased hygiene, reduced family size, vaccination, and consumption of increasingly hygienic foods in developed countries have altered the gut flora composition and consequently immune stimulation in early life. It has been proposed that these changes have contributed to the increase in allergic disease (68). Hence, there has been considerable research examining the observed benefits of probiotics to (i) affect the immune system (*in vitro* assessment), (ii) reduce atopic allergy in infants and adults, and also (iii) reduction of inflammation, especially with respect to inflammatory bowel diseases. Some proposed clinical functions are promotion of gut barrier function and enhanced intestinal IgA responses (69).

Researchers conducted a trial examining the modulating effect of probiotic milk intake on the humoral immune system (70). Their results indicated that lactic acid bacteria persisting in the gut can act as adjuvants to the humoral immune response. They observed that an increase in total serum IgA was observed in the test group compared to the control group.

Consumption of a probiotic milk drink containing *Bifidobacterium animalis* LKM512 by elderly, hospitalised patients was shown to ease constipation (71); this work in conjunction with further study (72) found that consumption of the probiotic milk drink contributed to suppressing inflammatory cytokine production of macrophages through production of anti-inflammatory metabolites *in vivo*.

2.1.4.8 *Effect of bifidobacteria on allergy in infants*

Rautava *et al.* investigated the effect of *Bifidobacterium* BB12 and *Lactobacillus* LGG in modulating the immune response in infants on introduction of cow's milk to their diet (73). Results indicated increased protective IgA responses to cows' milk compared with the control group, which was attributed to increased production of sCD14. Probiotic supplementation did not appear to stimulate development of cows' milk allergy, as none of the probiotic trial group developed cows' milk allergy, however 8% of the placebo group did.

There has also been interest in using probiotics to manage atopic eczema in infants. Researchers studied the impact of probiotic supplemented infant (hydrolysed whey formula) formula consumption in infants suffering from atopic eczema (69). Inflammation was measured via the SCORing Atopic Dermatitis (SCORAD) method (74), which characterises the severity of inflammation and assigns a number. After 2 months consumption, the SCORAD score for the trial group receiving *Bifidobacterium* BB12 supplemented formula dropped from >10 to 0 on average (range 0-3.8), compared with 13.4 for the control group. However, after 6 months consumption, all probiotic and control group SCORAD scores had dropped to an average of 0.

Fukushima *et al.* reported that infants fed infant formula containing *Bifidobacterium* BB12 for 21 days showed an increased level of fecal bifidobacteria, but also, increased fecal levels of total IgA and antipoliovirus IgA (75). It was postulated that these increases in IgA may contribute to enhancement of mucosal resistance against gastrointestinal infections; however, no direct link between IgA and BB12 consumption was examined.

Overall, probiotics appear to have some measurable capacity to modulate the infantile immune system in a positive manner. However, researchers still emphasise the need for further study to determine the mechanisms of these probiotic mediated effects, from two aspects: (i) the effect of the probiotic, and (ii) the reaction of the immune system (69, 73).

2.1.4.9 *Effect of bifidobacteria on cholesterol*

Hypercholesterolaemia is a major risk factor associated with cardiovascular disease, and reduction of blood serum cholesterol level is a desired factor in prevention of this disease. Consumption of fermented dairy products containing probiotic bacteria has been suggested as a means of reducing serum cholesterol (76, 77).

Several mechanisms for lowering of cholesterol by probiotics have been put forward, which include: cholesterol assimilation by the bacteria, cholesterol binding to the bacterial cell wall and enzymic deconjugation of bile acids, all of which prevent re-absorption of the bile acids into the body (78-80). However, the exact mechanisms involved have yet to be confirmed.

Pereira and Gibson studied the ability of several bacterial strains to assimilate cholesterol *in vitro* (81). Two strains of *B. infantis* were shown to reduce the cholesterol content of the broth medium studied. It was shown by Tahri *et al.* that some strains of bifidobacteria were able to reduce cholesterol *in vitro* through a combination of assimilation and precipitation of bile acids (82).

A rat-feeding study compared Bifidobacterial-supplemented yoghurt (soy and buffalo milks) , with unsupplemented control yoghurts, or unfermented soymilk

(83). The results showed significant increase in excreted fecal bile salt concentration and reduction of plasma and liver lipids in the bifidobacteria-supplemented yoghurt group, which exceeded the effects of the control groups (unsupplemented buffalo and soy milk yoghurts, and unfermented soy milk),. thus indicating that some strains of bifidobacteria have an effect in reducing cholesterol. Further study to confirm this effect in humans was advocated.

2.1.4.10 Effect of bifidobacteria on cancer

Colorectal cancer is one of the most common cancers in Western society, and accounts for approximately 20% of all cancers. It is in part due to genetic predisposition, but a diet low in fibre, and high in fat and protein has been linked to the aetiology of the disease (84), as this type of diet can result in increased levels of putrefactive organisms and a concomitant decrease in levels of bifidobacteria (85), which leads to an increase in fecal enzymes such as azoreductase, nitroreductase and urease, which have procarcinogenic activity.

There is growing evidence (conducted in animal models, or in tissue culture) that intestinal microorganisms can modulate carcinogenesis in the gut. This is achieved by prevention of colorectal tumour formation through several postulated effects.

Researchers examined the effect of administering *B. longum* one week after carcinogen exposure in mice, and found reduced aberrant crypt formation (ACF) in the gut compared to the control (86). Inclusion of inulin into the dosage mix increased the protective value of the *B. longum*. ACF is often used as a marker of precancerous tissue.

Commane *et al.* (87) studied the effect of probiotic fermentates with/without non-digestible oligosaccharides (NDOs) on the 'tight junctions' using the Caco-2 adenocarcinoma human cell line as a model. In the promotion phase of colorectal carcinogenesis, there is an observed loss in integrity across the gut mucosa, usually via weakening of the tight junctions, which bind the epithelial cells together and prevent the flow of solutes between cells. The strength of the tight junctions was measured using Trans-epithelial Electrical Resistance. (TER). This study found that exposure of Caco-2 cells to probiotic bacterial fermentates had varying effects on tight junction strength. A selection of lactobacilli and bifidobacteria were tested, and *B. animalis* BB12 had the greatest effect, especially when paired with Raftilose (NDO), increasing TER by 50% when bacterial fermentate alone was used, and up to 75% when BB12 was fermented with Raftilose. *Bif. Sp. 420* was a close second in terms of improved TER and tight junction integrity, and increased TER by 50% when combined with Raftilose. The authors postulated that bifidobacteria and lactic acid bacteria produce antiproliferative compounds during fermentation, and that selective addition of NDO can enhance this antiproliferative effect.

Leu *et al.* investigated the effect of a symbiotic combination of *B. lactis* and resistant starch on carcinogenesis in rats (88). They found that this combination facilitated the apoptotic response to a genotoxic carcinogen (azoxymethane-

AOM) in the distal colon of rats and offered increased protection against the early development of carcinogenesis.

Matsumoto and Benno investigated the effects of consumption of *B. lactis* LKM512-containing yoghurt on fecal probiotic metabolites and mutagenicity in seven healthy adults for 2 weeks (89). The fecal mutagenicity levels were significantly decreased during bifidobacterial yoghurt consumption and to a lesser extent during placebo consumption (yoghurt without Bifidobacteria). This was attributed to increased spermidine levels in the gut. Spermidine, along with polyamines and spermine has been reported to play a role in stabilising DNA, RNA, cell proliferation and regulation of enzymatic activity (90).

2.1.4.11 Effect of bifidobacteria on lactose intolerance

Approximately two thirds of the world's population suffer from lactose maldigestion, which is a biological result of a decline β-galactosidase in the gut at some point post-weaning. Therefore a person with lactose maldigestion lacks the ability to hydrolyse lactose in the small intestine, resulting in unhydrolysed lactose entering the colon. It is fermented by the indigenous flora into various gases and acidic fermentation products, resulting in symptoms such as abdominal pain and bloating, flatulence, and diarrhoea (91).

Several studies have looked at the potential of bifidobacteria to improve lactose digestion in lactose intolerant subjects. Jiang *et al.* (92) showed an improvement in lactose tolerance when milk containing *B. Longum* was consumed, compared with milk that had not been inoculated.

It is well documented that lactose consumed in the form of yoghurt is better tolerated than lactose from fresh milk (93). This may be attributed to the presence of *Streptococcus* spp. in yoghurt, which have a high β-galactosidase activity. They can reach the large intestine and thus ease negative fermentation effects. High β-galatosidase activity has also been demonstrated in streptococci *in vitro* (94).

A further study by He and colleagues showed an improvement in lactose tolerance amongst subjects that continued into the follow-up period, post-supplementation (95). Therefore, it may be possible to alleviate symptoms of lactose intolerance through modulation of the gut microflora in conjunction with dietary lactose (yoghurt) consumption.

2.1.5 Dosage

Depending on the means of administration, there is some variation in the cell counts of probiotic organisms used in various trials. A selection of dosage rates are given below in Table 2.1.I. The scientific literature appears to support a minimum probiotic dosage of 10^9 cfu/day, if a measurable benefit to the host is to be observed.

TABLE 2.1.I
Cell count delivery across a range of clinical studies.

Study	Dosage (cfu*/day)
Alander et al. (20)	3×10^{10}
Bartosch et al. (34)	7×10^{10}
Plummer et al. (49)	2×10^{10}
Wang et al. (55)	2.3×10^{9}
Sheu et al. (56)	4×10^{11}
Isolauri et al. (69)	$3\text{-}8 \times 10^{10}$
Link-Amster et al. (70)	3.75×10^{9}
Rautava et al. (73)	1×10^{10}
Orrhage et al. (96)	1×10^{11}

* cfu: Colony forming units

2.1.6   Safety

Bifidobacteria have been widely utilised in the food industry for over 20 years. Many researchers have observed their beneficial effects on the host and their ease of tolerance in clinical studies (55, 59, 60, 63), and some strains, such as Bifidobacterium BB12 have been awarded GRAS (Generally Regarded As Safe) status in the USA. However, as many researchers have indicated, the activity and suitability of each bacterial strain should be fully assessed individually, for both its safety and its probiotic efficacy (97, 98).

Within the European Community (EU), work is ongoing to assess various taxonomic groups of bacteria for Qualified Presumption of Safety (QPS) status, where bacteria will be assessed by EFSA (European Food Safety Authority) on the following criteria: established identity, body of knowledge, possible pathogenicity and end use. The reason behind this is to eliminate unnecessary and extensive work regarding safety assessment of bacteria already with a long history of safe use and an extensive body of scientific study supporting their safety. B. adolescentis, B. animalis, B. bifidum, B. breve and B. longum have all been put forward for QPS status (99).

A significant amount of data has accrued in area of probiotics in regard to bifidobacteria since the concept was first proposed by Tissier in 1900 (2) and Metchnikoff in 1907 (100). Recent research has highlighted the interest and need for effective and affordable means of counteracting negative aspects of gastrointestinal health. However, further work to elucidate the mode of action of probiotics and also further investigation into their potential benefits (7, 101-103) is needed.

2.1.7   References

1. FAO/WHO. Joint FAO/WHO expert consultation on evaluation of health and nutritional properties of probiotics in food. http://www.mesanders.com/prodefn.asp. 2001.

2. Tissier H. Recherches sur la flore intestinale des nourrissons (État normal et pathologique). *Thesis*, Ed. G. Carré et C. Maud, University of Paris (med.), Paris, France, 1900.

3. Tissier H. Traitement des infections intestinal per la méthode de la flore bacterienne de l'intestin. *Critical Reviews of the Society for Biology*, 1906, 60, 359-61.

4. Orla-Jensen S. La classifications des bactéries lactiques. *Lait*, 1924, 4, 468-74.

5. Bergey's Manual of Determinative Bacteriology, 8th edition, Ed. Buchanan R.E., Gibbons N. E.. The Williams and Wilkins Company, Baltimore, Maryland. 1974.

6. Garrity G.M., Bell J.A., Lilburn T.G. Taxonomic Outline of the Prokaryotes. *Bergey's Manual of Systematic Bacteriology, 2nd Edition*. Release 5.0, Springer-Verlag, New York. http://dx.doi.org/10.1007/bergeysoutline200405

7. Leahy S.C., Higgins D.G., Fitzgerald G.F., van Sinderen D. Getting better with Bifidobacteria. *Journal of Applied. Microbiology*, 2005, 98, 1303-15.

8. Masco L., Ventura M., Zink R., Huys G., Swings J. Polyphasic taxonomic analysis of *Bifidobacterium animalis* and *Bifidobacterium lactis* reveals relatedness at subspecies level: reclassification of *Bifidobacterium animalis* as *Bifidobacterium animalis* subsp. *animalis* subsp. nov. and *Bifidobacterium lactis* as *Bifidobacterium animalis* subsp. *lactis* subsp. nov. *International Journal of Systematic and Evolutionary Microbiology*, 2004, 54, 1137-43.

9. Felis G.E., Dellaglio, F. Taxonomy of Lactobailli and Bifidoabcteria. *Current. Issues in Intestinal Microbiology*, 2007, 8, 44-61.

10. Meile L., Ludwig U., Gut C., Kaufmann P., Dasen G., Wenger S., Teuber M.. *Bifidobacterium lactis* sp. nov., a moderately oxygen tolerant species isolated from fermented milk. *Systematic and Applied Microbiology*, 1997, 20, 20-57.

11. de Vries W., Stouthammer A.H. Fermentation of glucose, lactose, mannitol and xylose by Bifidobacteria. *Journal. of Bacteriology*, 1968, 96, 472-8.

12. Monnet V., Condon S., Cogan T.M., Gripon J.C. Metabolism of Starter Cultures. in *Dairy Starter Cultures*, Eds Cogan T.M., Accolas J.P. New York, VCH Publishers Inc 1996, 47-56.

13. Biavatti B., Vescovo M., Torriani S., Bottazzi V. Bifidobacteria: history, ecology, physiology and applications. *The Annals of Microbiology*, 2000, 50, 117-31.

14. Crittenden R.G., Martinex N.R., Playne M.J. Synthesis and utilisation of folate by yoghurt starter cultures and probiotic bacteria. *International Journal. of Food. Microbiology*, 2003, 80, 212-22.

15. Deguchi Y., Morishita T., Mutai M. Comparative studies on synthesis of water-soluble vitamins among human species of Bifidobacteria. *Agricultural and Biological Chemistry*, 1985, 49, 13-16.

16. Dong X., Xin Y., Jian W., Liu X., Ling D. *Bifidobacterium thermacidophilum* sp. Nov., isolated from an aerobic digester. *International Journal.of Systemic and Evolutionary Microbiology*, 2000, 50, 119-25.

17. Roberfroid M. Prebiotics: The Concept Revisited. *Journal of Nutrition*, 2007, 137, 830S-7S.

18. Vernazza C.L., Gibson G. R., Rastall R. A. Carbohydrate preference, acid tolerance and bile tolerance on five strains of Bifidobacterium. *Journal of Applied Microbiology, 2006, 100, 846-53.*

19. Macfarlance G.T., Steed H., Macfarlane S. Bacterial metabolism and health-related effects of galacto-oligosaccharides and other probiotics. *Journal.of Applied. Microbiology*, 2008, 104, 305-44.

20. Alander M., Mättö J., Kneifel W., Johansson M., Köggler B., Crittenden R., Mattila-Sandholm T., Saarela M. Effect of galacto-oligosaccharide supplementation on human fecal microflora and on survival and persistence of Bifidobacterium lactis Bb-12 in the gastro-intestinal tract. *International Dairy Journal*, 2001, 11, 817-25.

21. Tzortzis G, Goulas A.K., Gee J.M., Gibson G.R. A Novel Galactooligosaccharide Mixture Increases the Bifidobacterial Population Numbers in a Continuous *In Vitro* Fermentation System and in the Proximal Colonic Contents of Pigs *In Vivo. Journal of Nutrition*, 2005,135, 1726-31.

22. Waligora-Dupriet A-J., Campeotto F., Nicolis I, Bonet A, Soulaines P., Dupont C., Butel, M-J. Effect of oligofructose supplementation on gut microflora and well-being in young children attending a day care centre. *International Journal of Food Microbiology*, 2007, 113 (1), 108-13.

23. Tannock, G.W. Probiotic properties of lactic-acid bacteria: plenty of scope for fundamental R&D. *Trends in Biotechnology*, 1997, 15, 270-4.

24. Luckey T.D. Introduction to intestinal microecology. *American Journal of Clinical Nutrition*, 1972, 25, 1292-4.

25. Hopkins M.J., Sharp R., Macfarlane G.T. Age and disease related changes in intestinal bacterial populations assessed by cell culture, 16S rRNA abundance, and community cellular fatty acid profiles. *Gut*, 2001,48, 198-205.

26. Satokari R.M., Vaughan E.E., Smidt H., Saarela M., Matto J., de Vos W.M. Molecular approaches for the detection and identification of bifidobacteria and lactobacilli in the human gastrointestinal tract. *Systematic and Applied Microbiology*, 2003, 26, 572-84.

27. Benno Y., Sawada K., Mitsuoka T. The intestinal microflora of infants: Composition of fecal flora in breast-fed and bottle-fed infants. *Microbiology. and Immunology*, 1984, 28, 975-86.

28. Bezirtzoglou E., Maipa V., Chotoura N., Apazidou E., Tsiotsias A., Voidarou C., Kostakis D., Alexopoulos, A. Occurrence of Bifidobacterium in the intestine of newborns by fluorescence in situ hybridisation. *Comparative Immunology, Microbiology, & Infectious Diseases*, 2006, 29, 345-52.

29. Hopkins M.J., Macfarlane G.T., Furrie E., Fite A., Macfarlane S. Characterisation of intestinal bacteria in infant stools using real-time PCR and northern hybridisation analyses. *FEMS Microbiology Ecology*, 2005, 54, 77-85.

30. Favier C.F., Vaughan E.E., De Vos W.M., Akkermans A.D. Molecular monitoring of succession of bacterial communities in human neonates. *Applied Environmental Microbiology*, 2002, 68, 219-26.

31. Favier C.F., De Vos W.M., Akkermans A.D. Development of bacterial and bifidobacterial communities in faeces of newborn babies. *Anaerobe*, 2003, 9, 219-29.

32. Gavini F., Cayuela C., Antoine J.M., Lecoq C., Levfebre B., Membre J.M., Neut C. .Differences in the distribution of bifidobacterial and enterobacterial species in human faecal microflora of three different (children, adults, elderly) age groups. *Microbiology, Ecology, Health and Disease*, 2001, 13, 40-5.

33. Hopkins M.J., Macfarlane G.T. Changes in predominant bacterial populations in human faeces with age and with Clostridium difficile infection. *Journal of Medical Microbiology*, 2002, 51, 448-54.

34. Bartosch S., Fite A., Macfarlane G.T., McMurdo M.E.T. Characterization of bacterial communities in faeces from healthy elderly volunteers and hospitalized elderly patients by using real-time PCR and effects of antibiotic treatment on the fecal microbiota. *Applied Environmental Microbiology*, 2004, 70, 3575-81.

35. Ziemer C. J.,Gibson G.R. An overview of probiotics, probiotics and synbiotics in the functional food concept: perspectives and future strategies. *International Dairy Journal*, 1998,8, 473-9.

36. Grill J.P., Manginot-Dürr C., Schneider F., Ballongue J. Bifidobacteria and Probiotic Effects: Action of Bifidobacterium species on conjugated bile salts. *Current Microbiology*, 1995, 31, 23-7.

37. Perrin S., Grill J.P.,Schneider F. Effects of fructooligosaccharides and their monomeric components on bile salt resistance in three species of Bifidobacteria. *Journal of Applied Microbiology*, 2000, 88, 968-74.

38. Kirjavainen P., Ouwehand A., Isolauri E., Salminen S. The ability of probiotic bacteria to bind to human intestinal mucus. *FEMS Microbiology Letters*, 1998,.167, 185-9.

39. Haschke F., Wang W., Ping G., Varavithya W., Podhipak A., Rochat F., Link-Amster H., Pfeiffer A., Diallo-Ginstl E., Steenhout P. Clinical trials prove the safety and efficacy of the probiotic strain Bifidobacterium BB12 in follow-up formula and growing-up milks. *Monatsschrift für Kinderheilkunde*, 1998, (S1), 146, 26S-30S.

40. Ouwehand A.C., Isolauri E, Kirjavainen P.V., Tolkko S., Salminen S.J. The mucus binding effect of Bifidobacterium lactis BB12 is enhanced in the presence of *Lactobacillus* GG and *Lactobacillus delbrueckii* ssp. *bulgaricus*. *Letters in Applied Microbiology*, 2000, 30, 10-13.

41. Juntunen M., Kirjavainen P.V., Ouwehand A.C., Salminen S.J., Isolauri E. Adherence of probiotic bacteria to human mucus in healthy infants and during rotavirus infection. *Clinical and Diagnostic Laboratory Immunology*, 2001, 8, 293-6.

42. Schiffrin E.J., Rochat F., Link-Amster H., Aeschlimann J.M.,Donnet-Hughes A. Immunomodulation of human blood cells following the ingestion of lactic acid bacteria. *Journal of Dairy Science*, 1995, 78, 491-7.

43. de Vrese M., Marteau P.R. Probiotics and Prebiotics: Effects on Diarrhea. *Journal of Nutrition*, 2007, 137, 803S-11S.

44. Touré R., Kheadr E., Lacroix C., Moroni O., Fliss I. Production of antibacterial substances by bifidobacterial isolates from infant stool active against *Listeria moncytogenes*. *Journal of Applied Microbiology*, 2003, 95, 1058-69.

45. Silva A.M., Bambirra E.A., Oliveira A.L., Souza P.P., Gomes D.A, Vieira E.C., Nicoli J.R. Protective effect of bifidus milk on the experimental infection with *Salmonella enteriditis* ssp. typhimurium in conventional and gnotobiotic mice. *Journal of Applied Microbiology*, 1999, 86, 331-6.

46. Silva A.M., Barbosa F.H.F., Duarte R., Vieira L.Q., Arantes R.M.E., Niocli J.R. Effect of Bifidobacterium longum ingestion on experimental salmonellosis in mice. *Journal of Applied Microbiology*, 2003,.97, 29-37.

47. Asahara T., Kensuke S., Nomoto K., Hamabata T., Ozawa A., Takeda Y. Probiotic bifidobacteria protect mice from lethal infection with shiga toxin-producing *Escherichia coli* O157:H7. *Infection and Immunity*, 2004, 72, 2240-7.

48. Gagnon M., Kheadr E.E., Le Blay G., Fliss I. *In vitro* inhibition of Escherichia coli O157:H7 by bifidobacterial strains of human origin. *International Journal of Food Microbiology*, 2003,.92, 69-78.

49. Plummer S., Weaver M.A., Harris J.C., Dee P., Hunter J. *Clostridium difficile* pilot study: effects of probiotic supplementation on the incidence of *C. difficile* diarrhea. *International Journal of Microbiology*, 2004, 7, 59-62.

50. Colombel J.F., Corot A., Neut C.,Romond, C. Yoghurt with *Bifidobacterium longum* reduces erythromycin-induced gastrointestinal effects. *Lancet*, 1987,.2, 43.

51. Wagner R.D., Pierson C., Warner T., Dohnalek M., Farmer J., Roberts L., Hilty M., Balish E. Biotherapeutic effects of probiotic bacteria on candidiasis in immunodeficient mice. *Infection and Immununity*, 1997, 65, 4165-72.

52. Tomoda T., Nakano Y., Kageyama T. Variation of intestinal *Candida* of patients with leukaemia and the effect of *Lactobacillus* administration. *Japanese Journal of Medical Mycology*, 1983,.24, 356-8.

53. D'Souza A.L., Chakravarthi R., Cooke J., Bulpitt C.J. Probiotics in prevention of antibiotic associated diarrhoea: meta-analysis. *British Medical Journal*, 2002, 324, 1361.

54. Lesbros-Pantoflickova D., Corthésy-Theulaz I., Blum A.L. *Helicobacter pylori* and probiotics. *Journal of Nutrition*, 2007, 137, 812S-8S.

55. Wang K.-Y., Li S.-N., Liu C.-S., Perng D.-S., Wu D.-C., Jan C.-M., Lai C.-H., Wang T.-N., Wang W.-M. Effects of ingesting *Lactobacillus* and *Bifidobacterium* containing yogurt in subjects with colonised *Helicobacter pylori*. *American Journal of Clinical Nutrition*, 2004, 80, 737-41.

56. Sheu B.-S., Cheng H.-C, Kao A.-W., Wang S.-T., Yang Y.-J., Yang H.-B., Wu J.-J. Pretreatment with *Lactobacillus* and *Bifidobacterium* containing yogurt can improve the efficacy of quadruple therapy in eradicating residula *Helicobacter pylori* infection after failed triple therapy. *American Journal of Clinical Nutrition*, 2006, 83, 864-9.

57. Edwards C.A., Parrett A.M. Intestinal flora during the first months of life: new perspectives. *British Journal of Nutrition*, 2002, 88, Suppl.1, S11-S18.

58. Choraqui J.-P., Van Egroo L.-D., Fichot M.-C. Acidified milk formula supplemented with Bifidobacterium lactis: Impact on infant diarrhea in residential care settings. *Journal of Pediatric Gastroenterology And Nutrition*, 2004, 38, 288-92.

59. SaavedraJ.M., Bauman, N.A., Oung I., Perman J.A, Yolken R.H. Feeding of *Bifidobacterium bifidum* and *Streptococcus thermophilus* to infants in hospital for prevention of diarrhoea and shedding of rotavirus. *The Lancet*, 1994, 344, 1046-9.

60. Fukushima Y., Li S.-T., Hara H., Terda A., Mitsuoka T. Effect of follow-up formula containing Bifidobacteria (NAN BF) on fecal flora and fecal metabolites in healthy children. *Bioscience and Microflora*, 1997, 16, 65-72.

61. Ericsson C.D. Traveller's diarrhoea. *International Journal of Antimicrobial Agents*, 2003, 21,116-24.

62. Black F.T., Andersen P.L., Orskov J. Prophylactic efficacy of lactobacilli on traveller's diarrhea. *Travel Medicine*, 1989.7, 333-5.

63. Murray F.E., Bliss C.M. Geriatric constipation: a brief update on a common problem. *Geriatrics*, 1991, 46, 64-8.

64. Alm L., Ryd-Kjellen E., Setterberg G., Blomquist L. Effect of new fermented milk product, Cultura, on constipation in geriatric patients. 1st Lactic Acid Bacteria Norfolk, Computer Conf. Proc. Horizon Scientific Press, 1993.

65. Sagen O. Treatment of functional disturbance in the intestine by administration of lactic acid bacteria. Chr. Hansen A/S, Denmark. *Internal report*, 1989.

66. Tanaka R., Shimosaka K. Investigation of the stool frequency in elderly, who are bedridden and its improvement by ingesting bifidus yogurt. *Japanese Journal of Geriatrics*, 1982, 19, 577-82.

67. Burke W., Fesinmeyer M., Reed K., Hampson L., Carlsten C. Family history as a predictor of asthma risk. *American Journal of Preventative Medicine*, 2003, 24, 160-9.

68. Strachan D.P. Hay fever, hygiene and household size. *British Medical Journal*, 1989, 299, 1259-60.

69. Isolauri E., Arvola T., Sütas Y., Moilanen E, Salminen, S. Probiotics in the management of atopic eczema. *Clinical and Experimental Allergy*, 2000, 30, 1604-10.

70. Link-Amster H., Rochat F., Saudan K.Y., Mignot O., Aeschlimann J.M. Modulation of a specific humoral immune response and changes in intestinal flora mediated through fermented milk intake. *FEMS Immunology and Medical Microbiology*, 1994, 10, 55-64.

71. Matsumoto M., Imai T., Hironaka T., Kume H., Watanabe M., Benno, Y. Effect of yogurt with *Bifidobacterium lactis* LKM512 in improving fecal microflora and defecation of healthy volunteers. *Journal of Intestinal Microbiology*, 2001, 14, 97-102 .

72. Matsumoto M., Benno Y. Anti-flammatory metabolite production in the gut from the consumption of probiotic yogurt containing *Bifidobacterium animalis* subsp. *Lactis* LKM512. *Bioscience, Biotechnology and Biochemistry*, 2006, 70, 1287-92.

73. Rautava S., Arvilommi H., Isolauri E. Specific probiotics in enhancing maturation of IgA responses in formula fed infants. *Pediatric Research*, 2006, 60, 222-5.

74. European Task Force on Atopic Dermatitis. Severity scoring of atopic dermatitis. The SCORAD index. *Dermatology*, 1993, 186 (1), 23-31.

75. Fukushima Y., Kawata Y., Hara H., Terada A., Mitsuoka T. Effect of a probiotic formula on intestinal immunoglobulin A production in healthy children. *International Journal of Food Microbiology*, 1998, 42, 39-44.

76. Larsen L.A., Raben A., Haulrik N., Hansen A.S., Manders M., Astrop A. Effect of 8 week intake of probiotic milk products on risk factors for cardiovascular diseases. *European Journal of Clinical Nutrition*, 2000, 54, 288-97.

77. Bertolami M.C., Faludi A.A., Batlouni M. Evaluation of the effects of a new fermented milk product (Gaio) on primary hypercholesterolemia. European *Journal of Clinical Nutrition*, 1999, 53, 97-101.

78. Tahri K., Grill J.P.,Schneider F. Bifidobacteria strains behaviour toward cholesterol: coprecipitation with bile salts and assimilation. *Current Micribiology*, 1996, 33, 187-193.

79. Tahri K., Grill J.P.,Schneider F. Involvement of trihydroxyconjugated bile salts in cholesterol assimilation by Bifidobacteria. *Current Microbiology*, 1997, 34, 79-84.

80. Klaver F.A.M., van der Meer R. The assumed assimilation of cholesterol by lactobacilli and *Bifidobacterium bifidum* in due to their bile salt-deconjugating activity. *Applied and Environmental Microbiology*, 1993, 59, 1120-4

81. Pereira D.I.A., Gibson G.R. Cholesterol assimilation by lactic acid bacteria and Bifidobacteria isolated from the human gut. *Applied and Environmental Microbiology*, 2002, 68, 4689-93.

82. Tahri K., Crociani J., Ballongue J., Schneider F. Effects of three strains of Bifidobacteria on cholesterol. *Letters in Applied Microbiology*, 1995,.21, 149-51.

83. Abd El-Gawad I., El-Sayed E.M., Hafez S.A., El-Zeini H.M., Saleh F.A. The hypocholesterolaemic effect of milk yoghurt and soy-yoghurt containing Bifidobacteria in rats fed on a cholesterol-enriched diet. *International Dairy Journal*, 2005,15, 37-44.

84. Gill C.I.R.,Rowland I.R., Diet and cancer: assessing the risk. *British Journal of Nutrition*, 2002, 88, S1, S73-S87.

85. Benno, Y., Mitsuoka, T. & Kanazawa., K. 1991. Human fecal flora in health and colon cancer. *Acta Chirurgica Scandinavica*, S562, 15-23.

65. Rowland I., Rumney C., Coutts J., Lievense L. Effects of *Bifidobacterium longum* and inulin on gut bacterial metabolism and carcinogen induced aberrant crypt foci in rats. *Carcinogenesis*, 1998, 19, 281-5.

87. Commane D.M., Shortt C.T., Silvi S, Cresci A., Hughes R.M., Rowland I.R. Effects of fermentation products of pro- and probiotics on transepithelial electrical resistance in an *in vitro* model of the colon. *Nutrition and Cancer*, 2005, 51, 102-9.

88. Leu R. L., Brown I.L., Hu Y., Bird A.R., Jackson M, Esterman A.,Young G.P. A symbiotic combination of resistant starch and *Bifidobacterium lactis* facilitates apoptotic deletion of carcinogen-damaged cells in rat colon. *Journal of Nutrition*, 2005, 135, 996-1001.

89. Matsumoto M., Benno Y. Consumption of *Bifidobacterium lactis* LKM512 yogurt reduces gut mutagenicity by increasing gut polyamine contents in healthy adult subjects. *Mutation Research*, 2004, 568, 147-53.

90. Sreekumar O., Hosono A. The antimutagenic properties of a polysaccharide produced by *Bifidobacterium longum* and its cultured milk against some heterocyclic amines. *Canadian Journal of Microbiology*, 1998, 44, 1029-36.

91. Vesa T.H., Marteau P.,Korpela R. Lactose Intolerance. *Journal of the. American College of Nutrition*, 2000, 19, 165S-75S.

92. Jiang T., Mustapha A., Savaiano D.A. Improvement of lactoase digestion in humans by ingestion of unfermented milk containing *Bifidobacterium longum*. *Journal of Dairy Science*, 1996, 79, 750-7.

93. Marteau P., Flourie B, Pochart P., Chastang C., Desjeux J.-F., Rambaud J.-C. Effect of the microbial lactase (EC 3.2.1.23) activity in yogurt on the intestinal absorption of lactoase: an in vivo study in lactase-deficient humans. *British Journal of Nutrition*, 1990, 64, 71-9.

94. Sanders M.E., Walker D.C., Walker, K.M., Aoyama, K., Klaenhammer, T.R. Performance of commercial cultures in fluid milk applications. *Journal of Dairy Science*, 1996 79, 943-55.

95. He. T, Priebe M.G, Zhong Y., Huang C., Harmsen H.J.M., Raangs G.C., Antoine J-M.., Welling G.W.,Vonk R.J. 2008. Effects of yogurt and bifidobacteria supplementation on the colonic microbiota in lactose-intolerant subjects. *Journal of Applied Microbiology*, 2008, 104, 595-604.

96. Orrhage K., Sjöstedt S., Nord C.E. Effect of supplements with lactic acid bacteria and oligofructose on the intestinal microflora during administration of cefpodoxime proxetil. *Journal of Antimicrobial Chemotherapy*, 2000, 46, 603-11.

97. Fooks L.J.,Gibson G.R. Probiotics as modulators of the gut flora. *British Journal of Nutrition*, 2002, 88, S1, S39-S49.

98. Geier M.S., Butler R.N., Howarth G.S. Inflammatory bowel disease: Current insights into pathogenesis and new therapeutic options; probiotics, probiotics and synbiotics. *International Journal of Food Microbiology*, 2007, 115, 1-11.

99. Opinion of the Scientific Committee on a request from EFA on the introduction of a Qualified Presumption of Safety (QPS) approach for assessment of selected microorganisms referred to EFSA. Plus Appendix A - Assessment of gram-positive, non-sporulating bacteria. *The EFSA Journal*, 2007, 587, 1-16.

100. Metchnikoff E. *The prolongation of life.* London, Heinemann, 1907.

101. Sheil B., Shanahan F., O'Mahony L. Probiotic Effetcs on inflammatory bowel disease. *Journal of Nutrition*, 2007, 137, 819S-24S.

102. Shanahan F. Probiotics and inflammatory bowel disease: from fads and fantasy to facts and future. *British Journal of Nutrition*, 2002, 88, S1, S5-S9.

103. MarteauP.R., de Vrese M., Cellier C.J., Schrezenmeir J. Probiotics from gastrointestinal diseases with the use of probiotics. *American Journal of clinical Nutrition*, 2007, 73, 430S-6S.

2.2 LACTOBACILLI

Dr. Maria Saarela
VTT Technical Research Centre of Finland
PO Box 1000
FIN-02044 VTT
Finland

2.2.1   Description

Lactobacilli are Gram-positive non-sporeforming, catalase-negative (when growing without a heme source), usually nonmotile and typically nitrate-reducing rod-shaped bacteria. They utilise glucose fermentatively and may be either homofermentative - producing mainly lactic acid from glucose or heterofermentative - producing lactic acid, CO_2, ethanol and/or acetic acid. Bacteria of the genus *Carnobacterium*, which can be isolated from meat, poultry and fish, are phenotypically closely related to lactobacilli (1). At the time of the publication of the most recent edition of Prokaryotes in 2006 there were 80 *Lactobacillus* species. According to the most recent listing in 2008, the number of the species is around 120. The recently identified new *Lactobacillus* species originate from various habitats; sourdough, plant material, and human or animal gastrointestinal tract (GI) (2). The number of species will probably continue to change in the future: new species will still be identified, but some of the existing ones may be merged to form subspecies within a species, or transferred into another genera altogether (e.g. *Carnobacterium*, *Weissella*, *Atopobium*).

Based on phylogenetic relationship studies (using 16S rRNA gene sequences), the family Lactobacillaceae in the phylum Firmicutes contains the genera *Lactobacillus* and *Pediococcus*. Leuconostocaceae (with genera *Leuconostoc* and *Weissella*) is the most closely related family to Lactobacillaceae. Lactobacilli can be divided into several groups: *Lactobacillus buchneri* group, *Lactobacillus casei* group, *Lactobacillus delbrueckii* group, *Lactobacillus plantarum* group, *Lactobacillus reuteri* group, *Lactobacillus sakei* group and *Lactobacillus salivarius* group. The *L. buchneri* group contains species that can typically be found in food fermentations and spoilage. The *L. casei* and *L. delbrueckii* groups contain some of the important species for probiotic applications, the former *L. casei* and *Lactobacillus rhamnosus*, and the latter the whole earlier *Lactobacillus acidophilus* group (including *L. acidophilus* and *Lactobacillus johnsonii*). Both groups contain species from various habitats; e.g. the *L. delbrueckii* group bacteria can be isolated from human and animal sources as well as from food fermentation and spoilage. Other *Lactobacillus* species important in probiotic applications can be found in the groups *L. plantarum* (containing species mostly originating from plant/food material), *L. reuteri* (food, human, animal sources) and *L. salivarius* (mostly human and animal sources) (1, 2).

2.2.2   General properties

Although lactobacilli have complex nutritional requirements, they can be found in a wide variety of habitats, provided that their requirements for soluble carbohydrates, protein breakdown products and vitamins, and their need for a low oxygen tension are fulfilled. Lactobacilli are aciduric or acidophilic, and their production of organic acids (and concomitant lowering of the pH of their environment) gives them a competitive advantage against other microbes in the same habitat. Lactobacilli can be found in human and animal oro-gastrointestinal tracts, in plants and materials of plant origin, in soil, sewage and manure, and in food fermentations and spoilage (1). Many of the *Lactobacillus* species can typically be found in only one or few habitats, whereas some species are less specialised.

For probiotic applications, *Lactobacillus* species that can survive and potentially also colonise the GI tract are of interest. In the human oro-gastrointestinal tract lactobacilli can be found in the oral cavity, in the stomach and in the intestines. Lactobacilli have also been detected in the mucosal tissue of terminal ileum and colon (3). The oral cavity contains different micro-environments; cheeks, palate, tongue, tooth surfaces, gingival areas, and saliva have their own microbiota (4). Salivary microbiota reflects a mixture of bacteria washed off from the various surfaces, especially from the tongue (5). A large diversity of bacteria has been detected in the oral cavity and in the faeces, whereas the microbiota of the stomach is less complex. It has been shown with DNA-based techniques that although lactobacilli can be found in most subjects' faecal samples, they still comprise a small population among faecal bacteria in adults; <1-2 % of the total faecal bacteria (6, 7).

In the oral cavity lactobacilli have been commonly detected using culture techniques (8, 9, 10, 11), and also with real-time PCR (12, 13), whereas in several clone library studies , where low numbers of clones have been analysed, they have not been detected at all in various oral samples (14, 4, 15) One exception is the study of Paster *et al.*, where about 2,500 clones were analysed, and *Lactobacillus pontis*, *Lactobacillus casei* and *Lactobacillus brevis* were detected in the subgingival plaque samples (16). Culturable oral and faecal lactobacilli numbers vary greatly among subjects; among 10 Finnish volunteers studied by Maukonen, the cultured levels of oral lactobacilli were between 10^4 -10^6 cfu/ml of stimulated saliva, whereas in the faecal samples of the same subjects the lactobacilli levels were between 10^5-10^8 cfu/g (11). Similar levels have also been observed in other culture-based studies (17, 10). In the study by Maukonen, all subjects harboured culturable lactobacilli in the oral cavity and in the colon (11). Lactobacilli represented 0.1 and 0.01 % of the cultured bacteria in these sites, respectively. Three to six salivary genetic fingerprints (corresponding to strains) and 1-5 faecal genetic fingerprints were detected among the cultured (viable) lactobacilli in an individual. From stimulated saliva *Lactobacillus rhmanosus*, *L. casei/Lactobacillus paracasei*, *L. brevis*, *Lactobacillus gasseri*, *L. plantarum* gr. (*plantarum/arizonsis/pentosus*), *Lactobacillus jensenii*, *Lactobacillus parabuchneri*, *Lactobacillus fermentum*, *Lactobacillus vaginalis*, *L. salivarius* and *Lactobacillus kitasonis/Lactobacillus crispatus* were cultured, whereas from faeces *L. rhmanosus*, *L. paracasei*, *L. brevis*, *L. gasseri*, *L. plantarum* gr., *Lactobacillus ferintoshensis*, *L. crispatus* and *L. sakei* were isolated. Identical indigenous genetic fingerprints were found from saliva and faeces for eight of the ten subjects. These included *L. rhamnosus*, *L. gasseri*, *L. paracasei*, the *L. plantarum* group (comprising *L. plantarum*, *Lactobacillus arizonensis* and *Lactobacillus pentosus*) and a *Lactobacillus* sp., indicating that at least some *Lactobacillus* species/strains are able to live in different habitats in the human oro-gastrointestinal tract – although the oral cavity and colon differ in several aspects including e.g. redox potential, nutrients, mucosal surfaces, and co-existing members of the specific microbial community (11). The similarity of part of the faecal and salivary *Lactobacillus* populations has also been shown in further studies (18). In other studies some additional *Lactobacillus* species have been cultured from faecal samples, e.g. *L. acidophilus*, *L. salivarius*, *L. fermentum*, *L. buchneri* and *L. vaginalis* (19, 18). Several *Lactobacillus* species have been cultured from human stomach contents or mucosa including e.g. *L. fermentum*, *L. salivarius*, *L. plantarum*, *L. gasseri* and the novel species *Lactobacillus gastricus*, *Lactobacillus antri*, *Lactobacillus kalixensis* and *Lactobacillus ultunensis* (20, 21). Lactobacilli can also be found in the small intestine; in the upper jejunum the culturable numbers are around 100 cfu/ml (22).

Culture-based studies indicate that several *Lactobacillus* species can survive in the oro-gastrointestinal tract of humans. It is very likely that even more *Lactobacillus* species have been cultured from the human GI tract, but due to the taxonomic changes in the *Lactobacillus* genus only the most recent studies (with more up-to-date identification methods and taxonomy) have been included here. It has to be kept in mind that people ingest lactobacilli in their foods every day and therefore lactobacilli originating from the diet can potentially be detected in oral, stomach and intestinal samples. DNA-based detection enables the detection of dead bacterial cells: therefore the possibility of detecting transient bacteria is even greater than with culture-based studies. A repeated detection of a specific *Lactobacillus* species alive in a specific site probably indicates that this species is part of the indigenous microbiota of the site.

2.2.3   Applications

Typically, probiotics have been incorporated into dairy products, mainly fermented ones such as yoghurt, yoghurt drinks, cheese, quark, cultured buttermilk and dairy drinks. Other probiotic food products include milk, ice cream, fruit and berry juices and drinks, recovery drinks, cereal-based drinks and snacks. An early example of a probiotic dairy product is sweet acidophilus milk, which was introduced in the United States market as early as the 1970s (23) but gained in popularity among consumers much later. In the last few years the range of probiotic food products has increased considerably, since the food industry wants to offer alternative product options to consumers. Europe has traditionally had a strong position in the probiotic food market, which currently offers a wide range of probiotic food products (24, 25). Daily-dose dairy drinks have been the largest growing probiotic product type in the European market (26). In Europe *L. rhamnosus* GG (LGG) is perhaps the most diversely used probiotic in various foods (both dairy and non-dairy); LGG-containing foods are available in at least 15 European countries under various brand names (www.valio.fi). Other *Lactobacillus* probiotics available in foods in Europe include e.g. *L. casei* (strains Shirota, Immunitas, F19), *L. plantarum* 299v, *L. johnsonii* LJ1 (=LC1), *L. reuteri* and *L. acidophilus*.

Although the market for probiotic foods has expanded and developed substantially, surveys of probiotic products on the market today reveal quantitative and qualitative deficiencies particularly with respect to the labelling and viability of probiotic strain(s) (27, 28). Good viability is generally considered a prerequisite for optimal probiotic functionality (29) and therefore probiotic products should contain high enough levels of the specific probiotic strain(s) throughout the storage period and during consumption. Several factors such as strain characteristics, and food matrix properties, e.g. pH and temperature, such as temperature profiles during storage, and accompanying microbes, affect the viability and stability of a probiotic strain in the food product. In many food matrices, including beverages, the most important single factor affecting probiotic viability is probably the pH. Shelf-stable beverages typically have pH-values below 4.4 to ensure their microbial stability (30); e.g. fruit juices often have a pH below four or even three (31). A low pH combined with a long storage period is detrimental to the viability of most probiotic strains.

2.2.4   Physiological properties - probiotics and their health-effects

There is a growing demand for functional foods that improve the health and well-being of consumers. Due to rapidly aging populations, especially in the United States and Europe, the functional food sector will be even more condition-oriented in the future than today. In the adult population, combating lifestyle-related diseases, and in the elderly, the possibility of averting or delaying age-associated degenerative diseases are important targets for functional foods. Many consumer groups such as children and elderly, and those with impaired immune functions such as diabetes; can be prone to food-related diseases and GI disorders. The symptoms of many GI disorders and diseases can potentially be alleviated and prevented by consuming probiotic foods. *Lactobacillus* species tested for their probiotic potential include e.g. *L. acidophilus, L. reuteri, L. casei, L. johnsonii, L. plantarum* and *L. rhamnosus* (32).

The health-effects attributed to probiotics are diverse. For example, these include alleviation of lactose intolerance symptoms, treatment of viral and antibiotic-associated diarrhoea, reduction of symptoms of antibiotic treatment of *Helicobacter pylori*, alleviation of atopic dermatitis symptoms in children and lowering the risk of allergy in infancy, alleviation of symptoms of inflammatory bowel disease (IBD) and irritable bowel syndrome (IBS), and enhancing the immune response (33, 34). Probiotics are postulated to be effective especially in the cases where the condition is at least partially caused by GI tract microbiota imbalance (due to an exogenous pathogen or indigenous GI microbiota population shifts). However, the substantiation of many of the health claims attributed to probiotics is still not strong enough. According to the Food and Agriculture Organisation (FAO) food and nutrition paper (35) the claims related to prevention and treatment of acute diarrhoea in children and modulation of immune response in particular have good substantiation, whereas many other proposed health effects need to be further studied (Table 2.2.I). Many of the health effects are still debated in the literature and recent meta-analysis studies have sometimes given contradicting conclusions of probiotic efficacy. Thus, more high quality clinical trials are still needed to support probiotic health claims (see e.g. 36, 37, 38, 39, 40, 41, 42, 43, 44, 45).

TABLE 2.2.I
Health Benefits of probiotics (35)

Health benefit	Strong /fairly strong evidence	Some evidence	Note
Prevention of diarrhoea caused by certain pathogenic bacteria and viruses	X		evidence mainly from *Lactobacillus rhamnosus* GG and *Bifidobacterium animalis* subsp. *lactis* Bb-12
Activity against *H. pylori*		X	human data are still limited
Inflammatory diseases and bowel syndromes		X	human data are still limited
Anti-cancer effects		X	human data are still limited
Alleviation of constipation		X	human data are still limited
Immunomodulation (including allergy prevention)	X		evidence e.g. from *L. rhamnosus* GG and HN001 and *B. animalis* subsp. *lactis* Bb-12 and HN019
Beneficial effects in cardiovascular disease		X	human data are still limited
Lowering the risk of urinary tract infections recurrences	X		evidence e.g. from *Lactobacillus* sp. GR-1 and B-54
Eradication of bacterial vaginosis, and prevention and treatment of candidal vaginatis		X	human data are still limited

The FAO's food and nutrition paper (35) gives guidelines for the assessment of the properties of probiotics. The health benefits detailed in Table 2.2.I are strain-specific. These include issues such as: i) selection of probiotic strains for human use, i.e. probiotics must be able to exert their benefits on the host, but their source is not important; ii). classification and identification of individual strains (since probiotic properties are strain-related, identification has to be performed to a strain-level using appropriate molecular fingerprinting techniques): iii) defining and measuring the health benefits of probiotics, i.e. minimum daily amount required, duration of use, evidence for health benefits from *in vitro*, animal (when appropriate) and human studies; iv) safety considerations, i.e. mobile antibiotic resistance genes and safety of consumption in humans.

2.2.5   Safety of *Lactobacillus* probiotics

Like other members of the indigenous microbiota of the oro-gastrointestinal tract, lactobacilli occasionally cause opportunistic infections in humans. Indigenous lactobacilli have been connected to certain dental infections, bacteremia, endocarditis and to rare cases of other infections (46, 47). Since probiotic consumption typically involves the ingestion of large numbers of viable bacterial cells (daily dose can be e.g. 10^9 cfu) safety aspects of probiotic consumption are of utmost importance. Data on survival of the probiotics within the GI tract, their translocation and colonisation properties, and the fate of probiotic-derived components, is important for the safety evaluation of probiotic consumption. These characteristics are strain-specific; it is known for example, that there is great variation in the survival of probiotic strains through the GI tract (48). Various intrinsic properties of probiotics can be considered in probiotic safety evaluation. These include biogenic amine production, bile salt hydroxylase, D-versus L-lactate production, mucin degradation, enzymic activities (e.g. nitroreductase and β-glucuronidase), platelet aggregation activity, and colonisation and production of toxic metabolites (49, 50). However, the relevance of probiotic metabolic activities detected *in vitro* often remains speculative, since these activities are also present among the members of the indigenous GI microbiota (51). An additional safety consideration is the possible transferability of antibiotic resistance genes between probiotics and the members of the GI microbiota. Lactobacilli naturally display a range of antibiotic resistances (52), but in most cases antibiotic resistance is not of the transmissible type, and therefore it does not usually form a safety concern. However, since some *Lactobacillus* strains have been shown to contain acquired antibiotic resistance genes, antibiotic susceptibility testing should always be performed for strains aimed at probiotic applications (53, 54, 55, 56).

To date there are only a few relevant reports of probiotic bacterium being found in an infection: a case of liver abscess (57), a case of endocarditis (58), both in elderly subjects, and some cases of bacteremia and endocarditis in infants and children (59, 60, 61); for a review see Boyle (62). All these cases have occurred in patients who are immunocompromised; and chronically diseased or debilitated.

In Finland, *Lactobacillus* strains isolated from bacteremia have been systematically characterised for almost 20 years. Although the yearly consumption of probiotic products containing lactobacilli has increased, the incidence of lactobacillemia has not increased (63). The demographic and clinical characteristics of patients with *Lactobacillus* bacteremia and the *Lactobacillus* isolates were thoroughly analysed and characterised (64, 65). The extensive follow-up period of *Lactobacillus* bacteremia cases indicated that the risk of serious infection by one single probiotic strain is very low.

Probiotic products have been safely consumed in large quantities for a long time in Europe and in Japan without an indication that they could be generally harmful to consumers' health. However, the above findings indicate that sporadic localised infections in mainly immunocompromised patients may occur and show that no zero risk can be attributed to consumption of living microbes. It has to been noted that isolating a probiotic strain in an infection site does not necessarily mean that there is a direct cause-effect relationship between the isolate and the disease, because the growth of the probiotic strain in the site may be a secondary effect.

2.2.6 Analytical methods – species and strain-level identification of lactobacilli

Due to the somewhat confusing and constantly changing taxonomy of *Lactobacillus* species, special attention should be given to a proper and up-to-date species-level identification of probiotic *Lactobacillus* strains. Phenotypic tests such as API50 CH cannot be considered to be adequate for proper identification, but they can function as supplementary tests. Typically, *Lactobacillus* isolates are identified by (partial) 16S rRNA gene sequencing. GenBank (66) has a vast number (thousands) of *Lactobacillus* 16S rRNA gene sequences, which can be used as a reference in the identification. Sequence comparisons of the 16S-23S rRNA gene intergenic spacer regions have also been used for the identification of *Lactobacillus* isolates (67). However, GenBank has a much more limited (<400) collection of these *Lactobacillus* sequences at the moment, which might hamper proper species identification. Other potentially useful genes for *Lactobacillus* species identification (based on the gene sequence) include *hsp60*, *tuf*, *recA*, *rpoB* and *cpn60* (68).

Specific *Lactobacillus* strains are typically identified with various molecular fingerprinting techniques such as amplified ribosomal DNA restriction analysis (ARDRA), randomly amplified polymorphic DNA (RAPD), pulsed field gel electrophoresis (PFGE), or ribotyping (69). All these techniques necessitate the culture-based analysis of the samples and isolation of pure cultures of *Lactobacillus* isolates. Strain-specific quantitative PCR techniques have also been developed for the direct detection of probiotic *Lactobacillus* strains (70,71). However, these PCR methods can only give information about the absence or presence of the probiotic strain, not about its viability in the sample, e.g. in the GI tract. In GI samples this technique thus gives an indication about compliance (whether volunteers have consumed the probiotic product or not) but not about the strain's ability to survive the passage through the GI tract. This viability information is often crucial, and currently it can reliably be obtained with culture-based analysis only.

2.2.7   References

1. Hammes W.P., Hertel C. The genera *Lactobacillus and Carnobacterium, Prokaryotes*, 2006, 4, 320-403.

2. DSMZ, Bacterial nomenclature up-to-date:
   http://www.dsmz.de/download/bactnom/bactname.pdf

3. Ahmed S., Macfarlane G.T., Fite A., McBain A.J., Gilbert P., Macfarlane S. Mucosa-associated bacterial diversity in relation to human terminal ileum and colonic biopsy samples. *Applied and Environmental Microbiology*, 2007, 73, 7435-42.

4. Aas J.A., Paster B.J., Stokes L.N., Olsen I., Dewhirst F.E. Defining the normal bacterial flora of the oral cavity. *Journal of Clinical Microbiology*, 2005, 43, 5721-32.

5. Nisengard R.J., Newman M.G. *Oral microbiology and immunology*, Philadelphia, USA, W.B. Saunders Company, 1994.

6. Franks A.H., Harmsen H.J.M., Raangs G.C., Jansen G.J., Schut F., Welling G.W. Variations in bacterial populations in human feces measured by fluorescent *in situ* hybridization with group-specific 16S rRNA-targeted oligonucleotide probes. *Applied and Environmental Microbiology*,1998, 64, 3336-45.

7. Sghir A., Gramet G., Suau A., Rochet V., Pochart P., Doré J. Quantification of bacterial groups within human fecal flora by oligonucleotide probe hybridization. *Applied and Environmental Microbiology*, 2000, 66, 2263-6.

8. Rask P.I., Emilson C.G., Krasse B., Sundberg H. Dental caries and salivary and microbial conditions in 50-60-year-old persons. *Community Dentistry and Oral Epidemiology*, 1991, 19, 93-7.

9. Koll-Klais P., Mändar R., Leibur E., Marcotte H., Hammarström L., Mikelsaar M. Oral lactobacilli in chronic periodontitis and periodontal health: species composition and antimicrobial activity. *Oral Microbiology and Immunology*, 2005, 20, 354-61.

10. Caglar E., Kavaloglu S.C., Kuscu O.O., Sandalli N., Holgerson P.L., Twetman S. Effect of chewing gums containing xylitol or probiotic bacteria on salivary mutans streptococci and lactobacilli. *Clinical Oral Investigations*, 2007, 11, 425-9.

11. Maukonen J., Mättö J., Suihko M-L., Saarela M. Intraindividual diversity and similarity of salivary and fecal microbiota. *Journal of Medical Microbiology*, 2008, 1560-68.

12. Buyn R., Nadkari M.A., Chhour K-L., Martin F.E., Jacques N.A., Hunter N. Quantitative analysis of diverse *Lactobacillus* species present in advanced dental caries. *Journal of Clinical Microbiology*, 2004, 42, 3128-36.

13. Chhour K-L., Nadkarni M.A., Buyn R., Martin E., Jacques N.A., Hunter N. Molecular analysis of microbial diversity in advanced caries. *Journal of Clinical Microbiology*, 2005, 43, 843-9.

14. Kazor C.E., Mitchell P.M., Lee A.M., Stokes L.N., Loesche W.J., Dewhirts F.E., Paster B.J. Diversity of bacterial populations on the tongue dorsa of patients with halitosis and healthy patients. *Journal of Clinical Microbiology*, 2003, 41, 558-63.

15. Kang J-G., Kim S.H., Ahn T-Y. Bacterial diversity in the human saliva from different ages. *The Journal of Microbiology*, 2006, 44, 572-6.

16. Paster B.J., Boches S.K., Galvin J.L., Ericson R.E., Lau C.N., Levanos V.A., Sahasrabudhe A., Dewhirst F.E. Bacterial diversity in human subgingival plaque. *Journal of Bacteriology*, 2001, 183, 3770-83.

17. Mättö J., Fonden R., Tolvanen T., von Wright A., Vilpponen-Salmela T., Satokari R., Saarela M. Intestinal survival and persistence of probiotic *Lactobacillus* and *Bifidobacterium* strains administered in triple-strain yoghurt. *International Dairy Journal*, 2006, 16, 1174-80.

18. Dal Bello F., Hertel C. Oral cavity as natural reservoir for intestinal lactobacilli. *Systematic and Applied Microbiology*, 2006, 29, 69-76.

19. Mikelsaar M., Annuk H., Shchepetova J., Mänder R., Sepp E., Björksten B. Intestinal lactobacilli of Estonian and Swedish children. *Microbial Ecology in Health and Disease*, 2002, 14, 75-80.

20. Roos S., Engstrand L., Jonsson H. *Lactobacillus gastricus* sp. *nov.*, *Lactobacillus antri* sp. *nov.*, *Lactobacillus kalixensis* sp. *nov.* and *Lactobacillus ultunensis* sp. *nov.*, isolated from human stomach mucosa. *International Journal of Systematic and Evolutionary Microbiology*, 2005, 55, 77-82.

21. Wilson M. Bacteriology of humans. *An ecological perspective.* Oxford, UK, Blackwell Publishing Ltd, 2008.

22. Justesen T., Haagen Nielsen O., Jacobsen I.E., Lave J., Norby Rasmussen S. The normal cultivable microflora in upper jejunal fluid in healthy adults. *Scandinavian Journal of Gastroenterology*, 1984, 19, 279-82.

23. Tamime A.Y., Saarela M., Skriver A., Mistry V., Shah N.P. Production and maintaining viability of probiotic bacteria in dairy products, in *Probiotic Dairy Products*, Ed Tamime A.Y. , Oxford, UK, Blackwell Publishing, 2005, 39-72.

24. Halliwell D.E. Hands up for probiotics! The mechanism for health benefits is unclear, but interest remains strong. *The World of Food Ingredients*, 2002, 46-50.

25. Hilliam M. Healthier dairy (World Functional Dairy Products Market). *World of Food Ingredients*, 2004, 52-5.

26. Mellentin J. The key trends in functional foods 2006. *New Nutrition Business*, The Centre for Food & Health Studies, London, UK, 2006.

27. Coeuret V., Gueguen M., Vernoux J.P. Numbers and strains of lactobacilli in some probiotic products. *International Journal of Food Microbiology*, 2004, 97, 147-56.

28. Masco L., Huys G., De Brandt E., Temmerman R., Swings J. Culture-dependent and culture-independent qualitative analysis of probiotic products claimed to contain bifidobacteria. *International Journal of Food Microbiology*, 2005, 102, 221-30.

29. Saarela M., Mogensen G., Fonden R., Mättö J., Mattila-Sandholm T. Probiotic bacteria: safety, functional and technological properties. *Journal of Biotechnology*, 2000, 84, 197-215.

30. Eckert M., Riker P. Overcoming challenges in functional beverages. *Food Technology*, 2007, 3 (7), 20-6.

31. Saarela M., Virkajärvi I., Nohynek L., Vaari A., Mättö J. Fibres as carriers for *Lactobacillus rhamnosus* during freeze-drying and storage in apple juice and chocolate-coated breakfast cereals. *International Journal of Food Microbiology*, 2006, 112, 171-8.

32. Alvarez-Olmos M.I., Oberhelman R.A. Probiotic agents and infectious diseases: A modern perspective on a traditional therapy. *Clinical Infectious Diseases*, 2001, 32, 1567-76.

33. Reid G., Jass J., Sebulsky M.T., McCormick J.K., Potential uses of probiotics in clinical practice, *Clinical Microbiology Reviews*, 2003, 16, 658-72.

34. O'May G.A., Macfarlane G.T., Health claims associated with probiotics, in *Probiotic Dairy Products*, Ed Tamime A.Y., Oxford, UK, Blackwell Publishing, 2005, 138-66.

35. Anon. Probiotics in food – health and nutritional properties and guidelines for evaluation, *FAO Food and Nutrition Paper 85*, 2006.

36. Huang J.S., Bousvaros A., Lee J.W., Diaz A., Davidson E.J. Efficacy of probiotics use in acute diarrhea in children: a meta-analysis. *Digestive Diseases and Sciences*, 2002, 47, 2625-34.

37. Johnston B.C., Supina A.L, Vohra S. Probiotics for pediatric antibiotic-associated diarrhea: a meta-analysis of randomized placebo-controlled trials. *Canadian Medical Association Journal*, 2006, 175, 377-83.

38. Rolfe V.E., Fortun P.J., Hawkey C.J., Bath-Hextall F. Probiotics for maintenance of remission in Crohn's disease. *Cochrane Database of Systematic Reviews*, 2006, 18, 4.

39. Szajewska H., Ruszcynski M., Radzikowski A. Probiotics in the prevention of antibiotic-associated diarrhea in children: a meta-analysis of randomised controlled trials. *Journal of Pediatrics*, 2006, 149, 367-72.

40. McFarland L.V., Meta-analysis of probiotics for the prevention of traveler's diarrhea, *Travel Medicine and Infectious Disease*, 2007, 5, 97-105.

41. Takahashi O., Noguchi Y., Omata F., Tokuda Y., Fukui T. Probiotics in the prevention of traveler's diarrhea: meta-analysis. *Journal of Clinical Gastroenterology*, 2007, 41, 336-7.

42. Tong J.L., Ran Z.H., Shen J., Zhang C.X., Xiao S.D. Meta-analysis: the effect of supplementation with probiotics on eradication rates and adverse events during *Helicobacter pylori* eradication therapy. *Alimentary Pharmacology and Therapeutics*, 2007, 25, 155-68.

43. Elahi B., Nikfar S., Derakhshani S., Vafaie M., Abdollahi M. On the benefit of probiotics in the management of pouchitis in patients underwent ileal pouch anal anastomosis: A meta-analysis of controlled clinical trials. *Digestive Diseases and Sciences*, 2008, 53, 1278-84.

44. Huertas-Ceballos A., Logan S., Bennett C., Macarthur C. Dietary interventions for recurrent abdominal pain (RAP) and irritable bowel syndrome (IBS) in childhood. *Cochrane Database of Systematic Reviews*, 2008, 23, 1.

45. Lee J., Seto D., Bielory L. Meta-analysis of clinical trials of probiotics for prevention and treatment of pediatric atopic dermatitis. *Journal of Allergy and Clinical Immunology*, 2008, 121, 116-21.

46. Saarela M., Mättö J., Mattila-Sandholm T. Safety aspects of *Lactobacillus* and *Bifidobacterium* species originating from human oro-gastrointestinal tract or from probiotic products. *Microbial Ecology in Health and Disease*, 2002, 14, 233-40.

47. Cannon J.P, Lee T.A, Bolanos J.T, Danziger L.H. Pathogenic relevance of *Lactobacillus*: a retrospective review of over 200 cases. *European Journal of Clinical Microbiology and Infectious Disease*s, 2005, 24, 31-40.

48. Marteau P., Shanahan F. Basic aspects and pharmacology of probiotics: an overview of pharmacokinetics, mechanisms of action and side-effects. *Best Practice and Research in Clinical Gastroenterology*, 2003, 17, 725-40.

49. O'Brien J., Crittenden R., Ouwehand A.C., Salminen S. Safety evaluation of probiotics. *Trends in Food Science and Technology*, 1999, 10, 418-24.

50. Bernardeau M, Vernoux J.P., Henri-Dubernet S., Gueguen M. Safety assessment of dairy microorganisms: The *Lactobacillus* genus. *International Journal of Food Microbiology*, 2008, 126, 278-85.

51. Borriello S.P., Hammes W.P., Holzapfel W., Marteau P.M., Schrezenmeier J., Vaara M., Valtonen V. Safety of probiotics that contain lactobacilli or bifidobacteria. *Clinical Infectious Diseases*, 2003, 36, 775-80.

52. Danielsen M., Wind A. Susceptibility of *Lactobacillus* spp. to antimicrobial agents. *International Journal of Food Microbiology*, 2003, 82, 1-11.

53. Egervärn M., Danielsen M., Roos S., Lindmark H., Lindgren S. Antibiotic susceptibility profiles of *Lactobacillus reuteri* and *Lactobacillus fermentum*. *Journal of Food Protection*, 2007, 70, 412-18.

54. Klare I., Konstabel C., Werner G., Huys G., Vankerckhoven V., Kahlmeter G., Hildebrandt B., Müller-Bertling S., Witte W., Goossens H. Antimicrobial susceptibilities of *Lactobacillus*, *Pediococcus* and *Lactococcus* human isolates and cultures intended for probiotic or nutritional use. *Journal of Antimicrobial Chemotherapy*, 2007, 59, 900-12.

55. Huys G., D'Haene K., Danilesen M., Mättö J., Egervärn M., Vandamme P. Phenotypic and molecular assessment of antimicrobial resistance in *Lactobacillus paracasei* strains of food origin. *Journal of Food Protection*, 2008, 71, 339-44.

56. Ouoba L.I.I., Lei V., Jensen L.B. Resistance of potential probiotic lactic acid bacteria and bifidobacteria of African and European origin to antimicrobials: determination and transferability of the resistance genes to other bacteria. *International Journal of Food Microbiology*, 2008, 121, 217-24.

57. Rautio M., Jousimies-Somer H., Kauma H., Pietarinen I., Saxelin M., Tynkkynen S., Koskela M. Liver abscess due to a *Lactobacillus rhamnosus* strain indistinguishable from *L. rhamnosus* strain GG. *Clinical Infectious Diseases*, 1999, 28, 1160-1.

58. Mackay A.D., Taylor M.B., Kibbler C.C., Hamilton-Miller J.M.T. *Lactobacillus endocarditis* caused by a probiotic organism. *Clinical Microbiology and Infection*, 1999, 6, 290-2.

59. Kunz A.N., Noel J.M., Fairchok M.P. Two cases of *Lactobacillus* bacteremia during probiotic treatment of short gut syndrome. *Journal of Pediatric Gastroenterology and Nutrition*, 2004, 38, 457-8.

60. De Groote M.A., Frank D.N., Dowell E., Glode M.P., Pace N.R. *Lactobacillus rhamnosus* GG bacteremia associated with probiotic use in a child with short gut syndrome. *Pediatric Infectious Diseases*, 2005, 24, 278-80.

61. Land M.H., Rouster-Stevens K., Woods C.R, Cannon M.L. *Lactobacillus sepsis* associated with probiotic therapy. *Pediatrics*, 2005, 115, 178-81.

62. Boyle R.J., Robins-Browne R.M., Tang M.L.K. Probiotic use in clinical practice: what are the risks? *American Journal of Clinical Nutrition*, 2006, 83, 1256-64.

63. Salminen M.K., Tynkkynen S., Rautelin H., Saxelin M., Vaara M., Ruutu P., Sarna S., Valtonen V., Järvinen A. *Lactobacillus* bacteremia during a rapid increase in probiotic use of *Lactobacillus rhamnosus* GG in Finland. *Clinical Infectious Diseases*, 2002, 35, 1155-60.

64. Salminen M.K., Rautelin H., Tynkkynen S., Poussa T., Saxelin M., Valtonen V., Järvinen A. *Lactobacillus* bacteremia, clinical significance, and patient outcome, with special focus on probiotic *L. rhamnosus* GG. *Clinical Infectious Diseases*, 2004, 38, 62-9.

65. Salminen M.K., Rautelin H., Tynkkynen S., Poussa T., Saxelin M., Valtonen V., Järvinen A. *Lactobacillus* bacteremia, species identification, and antimicrobial susceptibility of 85 blood isolates. *Clinical Infectious Diseases*, 2006, 42, 35-44.

66. National Center for Biotechnology Information: http://www.ncbi.nlm.nih.gov

67. Tannock G.W., Tilsala-Timisjärvi A., Rodtong S., Ng J., Munro K., Alatossava T. Identification of *Lactobacillus* isolates from the gastrointestinal tract, silage, and yoghurt by 16S-23S rRNA gene intergenic spacer region sequence comparison. *Applied and Environmental Microbiology*, 1999, 65, 4264-7.

68. Blaiotta G., Fusco V., Ercolini D., Aponte M., Pepe O., Villani F. *Lactobacillus* strain diversity based on partial hsp60 gene sequences and design of PCR-restriction fragment length polymorphism assays for species identification and differentiation. *Applied and Environmental Microbiology*, 2008, 74, 208-15.

69. Satokari R., Vaughan E.E., Smidt H., Saarela M., Mättö J., de Vos W.M. Molecular approaches for the detection and identification of bifidobacteria and lactobacilli in the human gastrointestinal tract. *Systematic and Applied Microbiology*, 2003, 26, 572-84.

70. Coudeyras S., Marchandin H., Fajon C., Forestier C. Taxonomic and strain-specific identification of the probiotic strain *Lactobacillus rhamnosus* 35 within the *Lactobacillus casei* group. *Applied and Environmental Microbiology*, 2008, 74, 2679-89.

71. Fujimoto J., Matsuki T., Sasamoto M., Tomii Y., Watanabe K. Identification and quantification of *Lactobacillus casei* strain Shirota in human feces with strain-specific primers derived from randomly amplifies polymorphic DNA. *International Journal of Food Microbiology*, 2008, 126, 210-15.

3. SYNBIOTICS

Shelly Jardine
Leatherhead Food International
Randalls Road
Leatherhead
Surrey
KT22 7RY

3.1   Introduction

This chapter will introduce and describe synbiotics. Their general properties and their benefits to human health will be considered. It will also look briefly at the trend towards synbiotic usage in functional food products in the UK, Europe and USA.

The term *synbiotic* is increasingly found in literature on prebiotics and probiotics, with a number of ingredient manufacturers talking of the benefits of synbiotics. It is thought that the trend towards food products containing synbiotics will continue in the USA and Europe. There is, however, much work that needs to be done in the field of scientific research before consumers begin to understand and accept the term synbiotics.

3.2   Description

A number of definitions of synbiotics can be found in the literature. At the simplest level, a synbiotic is an ingredient that contains both a prebiotic and probiotic that work together to improve gut microflora. For the purpose of this chapter the following definition is used.

> Synbiotics are a mixture of prebiotics and probiotics that beneficially affect the host by improving the survival and implantation of live microbial dietary supplements in the gastrointestinal tract (1).

As synbiotics contain a mixture of prebiotics and probiotics it is also necessary to clearly define these.

'Prebiotic - is a non-digestible food ingredient that beneficially affects the host by selectively stimulating the growth and/or activity' of one or a limited number of bacteria in the colon, and this improves host health' (1).

'Probiotic – a live microorganism which when administered in adequate amounts confers a health benefit on the host' (2).

3.3   General properties

To understand how a synbiotic can potentially benefit human health it is necessary to briefly consider the properties of probiotics and prebiotics.

3.3.1   *Properties of probiotics*

The range of probiotic cultures available to the consumer has increased dramatically in the last 10 years. The bulk of commercial cultures for the food industry contain either *Lactobacillus* species or *Bifidobacterium* species. Lactobacilli belong to the Lactic Acid bacteria; Lactic Acid bacteria are Gram positive, non-mobile and non-spore forming (3).

The origin, habitat and species of *Lactobacillus* strains have a great impact on their value as probiotics (4). Over the past four decades there has been increasing interest in the isolation of novel *lactobacillus* strains that exert a beneficial health effect when ingested by humans (5). There have been a number of issues with the identification of probiotic strains, and the World Health Organisation (WHO) has now set out guidelines for the evaluation of probiotics in food (2). Probiotic bacteria display documented benefits not merely at a genus or species level, but at the level of a strain. The probiotic must reside in an internationally recognised cultural bank to allow for replication of current and future research (6).

The WHO probiotic criteria also apply to *Bifidobacterium*, the other common probiotic organism. *Bifidobacterium* are Gram positive rods, irregular and non-spore forming. Bifidobacteria are generally described as being strictly anaerobic. They do not grow well in milk even though most can ferment lactose (7).

Bifidobacterium are considered as key commensals in human-microbe interactions, and they are believed to play a pivotal role in maintaining a healthy gastrointestinal tract (GIT) (8). Numerous studies have shown that bifidobacteria constitute 95% of total gut bacteria in healthy breast-fed newborns (8). It is therefore not unsurprising that bifidobacteria are beneficial to humans in all stages of life.

The current synbiotic formulations that have either been tested, and/or are found in food products contain lactobacillus and/or bifidobacteria as well as a prebiotic.

3.3.2   *Properties of prebiotics*

The prebiotic group of ingredients are often identified in the research literature as "colonic foods". Prebiotics can pass through to the colon and become substrates for the host's bacteria. In order for a food ingredient to be classified as a prebiotic it must:

1. 'be neither hydrolysed nor absorbed in the upper part of the gastrointestinal tract ;

2. be a selective substrate for one or a limited number of beneficial bacterial comensal to the colon;

3. be able to alter the colonic flora in favour of a healthier composition, and;

4. induce luminal or systemic affects that are beneficial to the host health '(1)

The prebiotic ingredients covered under the above definitions include Lactulose (galactofructose), inulin and oligofructose, galacto-oligosaccharides. Importantly, prebiotics should only stimulate beneficial populations, and not the growth, pathogenicity or putrefactive activity of potentially deleterious microorganisms (9).

Therefore prebiotics act as a substrate for "good" gut bacteria. However in addition to modulating bacterial numbers, it is the effects of prebiotics on the metabolic activities of the microbiota that underpin many of the health benefits (10).

As with probiotics, a prebiotic affect must be demonstrated in humans to substantiate a claim (11). As such, considerable research has been undertaken on the various prebiotics and the amounts required to achieve a tangible health benefit. The required daily dosage for prebiotic effect varies from 4 g/day to 15 g/day (11). Another aspect of prebiotics to consider in a food product is their stability. Stability during the shelf-life is extremely important in order to ensure the desired dosage for a prebiotic effect at the end of a shelf life (12).

3.2.3   *Properties of synbiotics*

A synbiotic is a mixture of probiotics and prebiotics, and as such can potentially offer the properties of both probiotics and prebiotics. However, some researchers consider that synbiotics should be more than a mixture of the two ingredients; that a synergy must exist between the ingredients. This chapter, only reviews synbiotics that contain a prebiotic and probiotic, rather than focussing on the synergistic relationship between pre- and probiotic ingredients.

3.4   Physiological properties of synbiotics

The health benefits of using prebiotics and probiotics individually have been extensively researched, but there is much less information readily available on synbiotics. Synbiotics combine the documented benefits of probiotics with prebiotics, and their importance in the function of the gastrointestinal system, i.e.

the prebiotics support the probiotics to a greater extent than if the probiotic was the only functional ingredient.

The documented health benefits of probiotics are:

1. Lower frequency and duration of diarrhoea associated with antibiotics (*clostridium difficile*), rotavirus infection, chemotherapy and to a lesser extent travellers diarrhoea.

2. Stimulation of humoral and cellular immunity.

3. Decrease in unfavourable metabolites e.g. ammonium and precancerous enzymes in the colon (12).

Prebiotics are functional ingredients that stimulate the beneficial functional bacteria, primarily Lactobacilli and Bifidobacteria. Hence the associated health benefits are due to the growth or increase in growth of probiotics in the human gastrointestinal system. The scientific means by which prebiotics support probiotics is dealt with in previous chapters. One main health benefit to mention is the so called "bifidogenic effect" - food ingredients that are selectively fermented by colonic bacteria, leading to modification of the bacterial composition towards one that is predominated by bifidobacteria (13).

3.4.1 Digestive health

The potential benefit of synbiotics on human digestive health is of scientific interest and offers the potential for development of new functional foods. However, the current research on digestive health has primarily been focused on either prebiotics or probiotics separately. The research into digestive health covers areas such as treatment for serious conditions such as Inflammatory Bowel Disease (IBD), through to more common digestive problems for example, constipation and diarrhoea. Synbiotics as functional foods can affect different systems in the body, for example balance of colonic microflora, control of transit time and mucosal motility, and bowel habits (14).

Probiotics and prebiotics have been shown individually to have an effect on the treatment of diarrhoea. Probiotics are regularly utilised to treat antibiotic-associated diarrhoea which is often caused by the occurrence of *Clostridium difficile* after antibiotic treatment (15). Researchers are now looking at the potential benefits of synbiotics on different aspects of digestive health.

Research has now evaluated the benefits of synbiotics in lowering the level of infant diarrhoea and the safety of synbiotics for use in formula. Healthy infants were fed formulas containing *Bifidobacterium longum* BL999 + *Lactobacillus rhamnosus* + mixture of 90% galactooligosaccharides, and 10% fructooligosaccharide (16). This study confirmed the safety of different mixtures of probiotics and synbiotics (16). There was a potential decrease in the incidence of diarrhoea in the group given the synbiotic formula, an effect that lasted after the trial period; the authors recommended further investigation into the potential digestive health benefits of the synbiotic in treatment of diarrhoea.

Research evaluating toddlers consuming synbiotic formulas also noted similar decreases in diarrhoea incidence; the incidence of diarrhoea was lower for toddlers consuming the synbiotic formula, but the incidences were too low to draw definitive statistically significant conclusions (17).

A large research study conducted by Fisberg *et al.* evaluated the benefits of synbiotics on a large group of preschool children in decreasing sick days and incidence of diarrhoea. This study utilised *Bifidobacterium* and *Lactobacillus acidophilus* together with fructooligosaccharide in a formula. The synbiotic group experienced a significant reduction in constipation amongst all ages, and significantly reduced sick days among 3 to 5 year old children (18).

Synbiotics could also benefit patients suffering from diarrhoea associated with antibiotic treatment. Wensus (2007) reported a significant reduction in the risk of antibiotic-associated diarrhoea when patients were administered a multi-strain synbiotic treatment containing Müller *Bifidobacteria* BB-12 (19). Consumption of a probiotic yoghurt significantly reduced the duration of diarrhoea from 10 days to 4 days during *Heliobacter pylori* eradication (19). This research has supported the development of a synbiotic yoghurt drink; Müller's Activa®, which contains *Bifidobacterium* BB-12 and the prebiotic inulin. The next logical step is to look at synbiotic products like Müllers Activa® and others to aid in diarrhoea treatment to improve digestive health. Antibiotic-associated diarrhoea is often caused by *C.difficile*, and probiotics have been proven to be effective against this condition (20).

Another aspect of digestive health is constipation. Synbiotics developed for the prevention of constipation are now found in various yoghurt drinks. The probiotic strains vary, and the prebiotic is typically inulin. Recent findings indicate that the efficiency of probiotics in improving gut health is enhanced when probiotics are combined with dietary fibres (19).

Other serious aspects of digestive health are IBD and irritable bowel syndrome. There is evidence showing that the microbiota of patients with IBD differs from that of healthy patients (21). Several studies have shown that probiotics might have a beneficial effect in IBD patients (22, 23). Research now needs to be undertaken on the potential synbiotics for the alleviation of IBD.

An aspect of IBD is ulcerative colitis. One research paper looked at synbiotic therapy on patients with ulcerative colitis, and found that short-term synbiotic treatment of active ulcerative colitis resulted in the improvement of the full clinical appearance of chronic inflammation in patients receiving the therapy (24). The synbiotic contained *Bifidobacterium longum* and inulin-oligofructose. However, the study group was only 12 patients, with just six taking the synbiotic.

3.4.2 *Modulation of the immune system*

Numerous studies on *in vitro* models and rat studies have shown that probiotics and prebiotics individually have potential benefit on the human immune system.

Probiotic cultures, via their intimate association with the intestinal mucosa, their cellular components, and their effects on associated microbiota appear to improve immunological function in the GIT (25). This is because the indigenous intestinal microflora is a principal component of the intestine's defence barrier against antigens from microorganisms and food.

To further understand the complex immune reactions in the GIT it is best to read the research in detail. Ouwehand showed that adhesion of probiotics to mucosal surfaces leads to immune modulation, competitive exclusion of pathogens, and prevention of pathogen and transient colonisation (26). The GIT mucosal surfaces are, by the process of digestion, continually exposed to antigenic components, and the health of the human is partly ensured by the GIT mucosal immune response and the systemic immune response.

In the GIT there are specialised gut-associated lymphoid tissues (GALT) which include tonsils, adenoids, peyers patches, the appendix and solitary lymphoid nodules (27). Protection against certain infections can be transferred by serum, this is called humoral immunity (28). The immune reaction is mediated by circulating antibodies, known as immunoglobulins (Ig) (28). There are five immunoglobulins with different functionailities IgA, IgD, IgE, IgG and IgM. Immunoglobulins are major components of the humoral immune response system (28). IgA circulates in the bloodstream but of more functional importance, IgA is secreted across mucous membranes and is found in intestinal and bronchial secretions (28). A large amount of research on probiotics and immune modulation has concentrated on the effects on the humoral immune response system and the function of IgA (29).

It has also been shown in research that probiotics, stimulated by prebiotic fermentation are important in the development of sustainable intestinal defences; i.e. probiotics can stimulate the synthesis and secretion of polymeric IgA, the antibody that coats and protects mucosal surfaces against harmful bacteria (30). Research has shown that the probiotic effect of immune system modulation is due to the strengthening of non-specific and antigen-specific defences against infections and tumours, and an adjuvant effect on antigen-specific immune responses (31).

The mechanisms by which probiotic cultures appear to carry out beneficial activities in the GIT via probiotic and/ or abiotic mechanisms is detailed in Figure 3.1 on the following page.

When looking to improve immunity in patients that are suffering from chronic conditions, medical research is starting to consider the use of synbiotics as an adjunct to current antibiotic and invasive treatments. It is well known that antibiotics are detrimental to the human microbiota in the GIT. Often in patients with severe respiratory diseases low levels of *Bifidobacterium* and *Lactobacillus* were detected, and levels of pathogenic organisms were high (30). Clinical studies on synbiotics have shown that the altered bacterial flora that was a consequence of synbiotic treatment was accompanied by a clear improvement in the clinical condition of the patients (31, 32).

The potential for synbiotics to modulate immunity in humans seems promising. Research suggests that synbiotics can benefit the immune system and in particular, seem to activate the GIT immune system. However, further clinical studies are required that evaluate one clearly identified probiotic strain that aids in immunity in combination with a prebiotic.

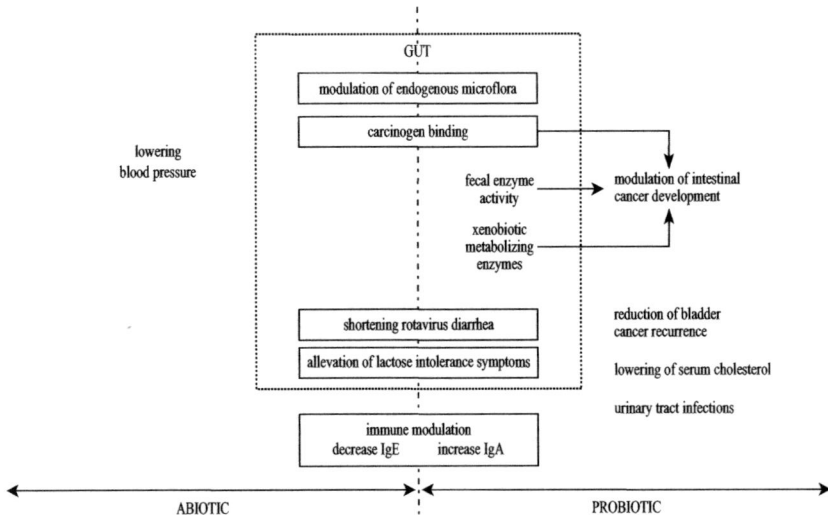

Fig. 3.1. Health benefits and suspected mechanisms of probiotics versus abiotics. IgE, IgA, immunoglobulins E and A

3.4.3   *Anti-carcinogenicity*

There are many different cancers that can affect the human body It was originally hypothesised that certain food components could be potentially dangerous to health. It is now well established that certain chemicals such as nitrates, in excess, are carcinogenic. However, research has indicated that consumption of certain foods or food components can help lower the risk of developing cancer. There is epidemiological evidence, that indicates a negative correlation between the incidence of certain cancers and the consumption of fermented milk (33). It is postulated that research into synbiotics will eventually reveal anti-carcinogenic effects.

Research on probiotics and anti-carcinogenicity has previously been carried out in the USA, Europe and Japan. The difficulty in summarising the research is

121

that the potential benefits are strain-specific. A large amount of research on the effects of synbiotics and probiotics on cancer has been performed in rat models looking at specific tumours and carcinogens. However, some trials do not utilise controls or placebo tests, limiting the utilisation of this research. This was noted by Ouwehand et al. who proposed that clinically validated trials should be double blind, placebo controlled and randomised, detailing the minimum dose for each strain (34).

There is a need to establish how synbiotic foods can potentially protect humans from cancer. Colon cancer is rated as a significant cancer by the World Health Organisation (WHO) and causes the deaths of 677,000 people a year (35). The European Union (EU) has sponsored a study of synbiotics and their effect on colon cancer. The SYNCAN project brings together scientists from six countries to test the hypothesis that prebiotics and probiotics in combination (synbiotics) protect the gut from DNA changes that trigger colon cancer (36).

The initial trials from the SYNCAN project looked at two systems; *in vitro* fermentation studies, and *in vivo* animal studies. The *in vivo* animal study was designed to compare the effects of the probiotics, the prebiotics and the synbiotic combination in a model of carcinogenesis (37).

The *in vivo* study results show that the prebiotic and synbiotic diets, but not the probiotic diet, significantly reduced the tumour incidence in the colon (36). Looking at whether the synbiotic actually showed greater benefits than just the prebiotic or probiotic alone is a research challenge. In the *in vivo* animal trial, the synbiotic-fed group had fewer tumour-bearing rats than the prebiotic only, and probiotic only groups (37).

Human studies have also been undertaken under the SYNCAN project, one study looked at patients treated for colon cancer as well as healthy subjects who had intestinal polypectomised (polyp) treatment. The synbiotic group received capsules of *Lactobacillus rhamnosus* GG and *Bifidobacterium lactis* Bb12, and 10 g of inulin enriched with oligofructose (38). After 12 weeks, the subjects in the synbiotic group saw reductions in colon cancer markers, normalisation of cell turnover and 60% decrease in mucosal DNA damage, with risk reduction strong in the polyp patients (36). Authors found that supplementation with synbiotics had minor stimulatory effects on the systemic immune system of the two study groups (38). It was also noted that IFN-y secretion was only affected after intake of the synbiotic for 12 weeks, suggesting that only long-term intake results in immunodulation (38). The authors suggested that further research was needed to focus specifically on the effects of synbiotics on the gut-associated immune system.

A summary of the SYNCAN projects and other research undertaken on synbiotics is presented in Table 3.I

TABLE 3.I
Selected Human Trials on Synbiotics

Synbiotic	Method	Result	Study group	Reference
L.acidophilus and . Bifidobacterium spp and fructooligosaccharide	626 healthy malnourished children	Decrease in number of sick days and constipation in children in synbiotic group	Control vs synbiotic	(Fisberg et al. 2002) (18)
B.lactis and fructo-oligosaccharide 1.5g	318 healthy toddlers	All formulas found to be safe, potentially the synbiotic lowered incidence of diarrhoea	Control vs probiotic vs synbiotic	(Bettler et al. 2006) (17)
B. longum BL999 and L.rhamnossus LPR and galactooligosaccharide + fructooligosaccharide	284 healthy infants	Formulas containing probiotics and synbiotics led to healthy weight gains	Control vs probiotic vs synbiotic	(Chouraqui et al. 2008) (16)
B. bifidum and B.lactis + 6g oligofructose	18 healthy humans	Increase in numbers of bifidobacteria in synbiotic group	Control vs synbiotic	(Bartosch et al. 2005) (39)
L.rhamnosus and B.lactisand inulin, oligofructose	37 colon patients, 43 poly-pectomised patients	Supplementation with synbiotics had minor stimulatory effects on the systemic immune system.	Control vs. synbiotic	(Roller et al. 2007) (38)
B.longum and inulin-oligofructose	18 ulcerative patients	Improvement in clinical appearance of chronic inflammation	Control vs. synbiotic	(Furrie et al 2005) (24)
L.paracasei and fructooligosaccharide 6g	60 elderly persons	Decrease in respiratory tract infections in the synbiotic group	Control vs. synbiotic	(Bunout et al. 2004) (40)
B.breve and L.casei and galactooligosaccharide	7 patients with severe respiratory distress	Improved bacterial flora, reduced risk of bacteremia and frequency of antibiotics	Synbiotic	(Kanamori et al. 2006) (31)

3.5   Applications

Synbiotic food is a relatively new category in the area of functional foods. In comparison, probiotic foods are well established functional products readily available to consumers primarily in a dairy format. Probiotics are commonly added to fermented dairy products and their quality in terms of viable counts and taxonomy has improved over the years (41). Scientific research on probiotics in food has been ongoing for over two decades or even longer if we consider

Metchnikoff's ground breaking research one hundred years ago (42). This research on probiotics is well documented and now specifically carried out at strain level. Probiotics are typically found in dairy-based products such as cheese and yoghurts. As technological developments have brought improvements in stability, probiotics are also finding applications in shelf-stable formats, for example breakfast cereals and confectionery products.

Prebiotic ingredients are now used in a broad range of consumer goods from breakfast cereals to biscuits, bread and cereal bars. Prebiotics are also added to fruit juices, fruit drinks, water, confectionery and dairy-based products.

Functional probiotic foods targeted towards improving the balance and activity of the intestinal microbiota, currently provide the largest segment of the functional food market in Europe, Japan and Australia (43). There now exists the opportunity to combine the benefits of probiotics and prebiotics into a range of synbiotic functional foods.

The suppliers of prebiotics and probiotics are now devoting increasing levels of research to identifying combinations of pre- and probiotics that have proven health benefits (44). Functional foods containing probiotics in the form of daily dose drinks or yoghurts are a well established category. Therefore this food format probably offers the greatest potential for new synbiotic products.

A recently released synbiotic is Müller's Vitality®, which contains a combination of prebiotic inulin and probiotic *Bifidobacterium* BB12®. The benefits of the synbiotic combination are reported as: improving intestinal health, helping replenish levels of bifidobacteria after antibiotics, and playing a role in reducing infection by *Clostridium difficile* (18).

Another synbiotic drink that is marketed as a functional food to improve digestive health is Yo-plus®. Yo-plus® is a new synbiotic drink that explains the prebiotic component to consumers as "natural fibre" to aid in consumer understanding of the prebiotic's function (45). The Yo-plus® synbiotic contains *Bifidobacterium* Bb12 and inulin. Yo-plus® is manufactured by General Mills and is available in the USA.

In the first 9 months of 2008 there were 142 new foods or beverages launched on the US Market in the digestive health category (46). This functional food category is growing rapidly and provides opportunities for further synbiotic functional foods. Detailed in the Table 3.II are some of the synbiotic products available in the market place.

TABLE 3.II
Synbiotic product examples

Product (Category)	Manufacturer	Country	Year of launch
WellB (Yoghurt and drinking yoghurt)	Parmalat	South Africa	2006
ABC (Children's yoghurt)	Central Lechera Asturiana	Spain	2007
Simbi (Milk drink)	Priegola	Spain	2003
Yomo Rinforzo Plus	Granarolo	Italy	2008
Müller Vitality (Yoghurt and yoghurt drink)	Müller dairies	UK	2008 (re-launch)
Activia Fibre (Yoghurt)	Danone	UK	2005
Denge (Yoghurt)	Pinar	Turkey	2005
Elivae (Yoghurt)	Yoplait/National Foods	Australia	2007
Yo-plus® (Yoghurt)	General Mills	USA	2007
Minigo Junior (Fromage frais)	Yoplait	Canada	2008
Digestive Health (Cottage cheese)	Friendship Dairies	USA	2008
LiveActive (Low fat cottage cheese)	Kraft Foods	USA	2007

Products containing synbiotics currently on the market focus on the functional benefits for gut health. Many of these foods contain functional microbes or ingredients that provide a health benefit in the human gut (43). The mechanisms by which this occurs are less well understood. Further research is required to discover the mechanisms behind health effects at a molecular level (43).

For the development of synbiotic functional food products the problem (and potential benefit) is the range of combinations of the various probiotics (large numbers of strains with different clinical benefits) with a range of prebiotics. *In vitro* tests have been conducted to determine the functional activity of various prebiotic carbohydrates combined with some probiotic strains (44). Further large clinically validated trials are required to establish proven benefits to consumers. Ouwehand *et al.* noted that in the area of establishing health benefits, considerable work remains. Fermented dairy products remain the most likely consumer products for supplementation with synbiotics (34).

The research of novel formulations with newly selected probiotic strains is important to satisfy the increasing demands of the market and to obtain functional products in which the probiotic cultures are more active and protected from gastrointestinal stress (47). The combination of prebiotics with probiotics seems to offer and confer stability to the probiotics, so that they are better able to withstand digestion and pass through to the colon. It is also important to provide quality assurance for the product and the probiotic strains used, to ensure that the properties of each strain are stable and do not alter during long periods of industrial utilisation (48).

So, when a manufacturer is investigating a potential new synbiotic, both the probiotic (strain and stability) and prebiotic have to be considered. This is particularly relevant for acidic foods that are processed at elevated temperatures (49). But increasingly synbiotics are being found in a range of different food formats.

Synbiotics are now being added to products such as chocolate mousse. Apart from potentially creating a functional food, the synbiotic combination can have positive sensory and textural benefits to the food. A study using inulin with *Lactobacillus paracasei* was found to have a favourable texture and sensory property (50). The synbiotic combination allowed for a firmer mousse, and authors noted that *in vivo* trials are now needed (50).

There is an opportunity for new synbiotic products to be developed in foods aimed at different age groups. One area to consider is the growing demographic of the elderly, this "Senior" market offers plenty of potential for synbiotics with proven benefits. Some research studies such as Bunout *et al.* have shown that nutritional supplementation has proven health benefits. This study showed a special nutritional formula (synbiotics and vitamins) given to the elderly had positive effects on the immune response, and a reduction in the number of infections (40).

Currently the food industry has many challenges in the functional food sector in ensuring consumer acceptance and understanding of products' health benefits, and achieving health authority approval of that functional food. For developers considering a synbiotic product, detailed guidelines on evaluating the probiotic component of the synbiotic can be found in the WHO 'Guidelines for the evaluation of probiotics in food' (2). Evaluation of commercially available prebiotics will be a challenge for the developer but it has been found in general only heating at low pH caused a significant reduction in prebiotic activity (51), developers are recommended to read Huebner *et al.* studies on prebiotics (51).

When developing a synbiotic health product it is important that suppliers and manufacturers look at substantiating their health claims. Often these require human intervention studies, which are time consuming and costly. Any health claim should be backed with solid scientific evidence (clinical trials, *in vitro* laboratory studies, and epidemiological studies) (52). The considerations to be given to the study design should include the dosages of the probiotic and prebiotic, frequency of administration, type of food for administration, duration of study, and how many groups to include (control, synbiotic, prebiotic and/or probiotic) (53).

3.6   References

1. Gibson G.R., Roberfroid, M.B. Dietary modulation of the human colonic microbiota: introducing the concept of Prebiotics. *Journal of Nutrition*, 1995, 125, 1401-12.

2. Food and Agriculture Organisation, World Health Organisation. *Guidelines for the evaluation of probiotics in food.* Guidelines, Ontario, Canada, May 2002 Ed. Food and Agriculture Organisation, World Health Organisation Geneva, WHO, 2002.

3. Walsta P., Wouters J.T, Geurts T.J. Lactic fermentations, in *Dairy Science and Technology 2nd Ed*, Eds Walsta P., Wouters J.T, Geurts T.J. CRC Press Boca Raton, Florida, 2006, 357-97.

4. Annuk H., Shchepetova J., Kullisar T., Songisepp E., Zilmer M., Mikelsaar M. Characterisation of intestinal lactobacilli as putative Probiotic candidates. *Journal of Applied Microbiology*, 2003, 94, 403-12.

5. Mabrouk A., Effat B., Sadek Z., Hussein G.A.M., Magdoub M.I.N. Probiotic properties of some lactobacillus strains. *International Journal of Probiotics and Prebiotics*, 2007, 2 (4) 175-84.

6. Decker K.J. The Science behind the Success of Probiotics and Prebiotics. *Food Product Design*, 2008, 18, 76-85.

7. Mogensen G., Skaaning T., Grenov B. Probiotics, in *Guide to Functional Food Ingredients*. Ed. Young J., Leatherhead, United Kingdom, Leatherhead Publishing. 2001, 20-50.

8. Leahy S.C., Higgins D.G., Fitzgerald G.F., van Sinderen D. Getting better with bifidobacteria. *Journal of Applied Microbiology*, 2005, 98, 1303-15.

9. Crittenden R. Emerging prebiotic carbohydrates, in *Prebiotics: Development & Application*, England, John Wiley & Sons Ltd., 2006, 11-134.

10. Roberfroid M.B. General introduction: Prebiotics in Nutrition, in *Handbook of Prebiotics*, Eds. Gibson G.R., Roberfroid M.B., Boca Raton, Florida USA, CRC Press, 2008, 1-12.

11. Sveje M. A new era for gut health. *Nutracos*, 2008, 7, 8-11.

12. Schrezenmeir J., de Vrese M. Probiotics, prebiotics, and synbiotics – approaching a definition. *American Journal of Clinical Nutrition*, 2001, 73, 362S-364S.

13. Franck A., Coussement P. Prebiotics, in *Guide to Functional Food Ingredients*, Ed Young J. Leatherhead, United Kingdom, Leatherhead Publishing, 2001, 1-19.

14. Nagpal R., Yadav H., Puniya A.K., Signh K., Jain S., Marotta F. Potential of Probiotics and Prebiotics for Synbiotic Functional Dairy Foods: An Overview. *International Journal of Probiotics and Prebiotics*, 2007, 2, 2/3 75-84.

15. Shah N.P., Functional cultures and health benefits. *International Dairy Journal*, 2007, 17, 1262-77.

16. Chouraqui J.M., Grathwall D., Labaune J.M., Hascoet J.M., Montgolfer I., de leclaire M., Giarre M., Steenhout P. Assessment of the safety, tolerance, and protective effect against diarrhoea of infant formulas containing mixtures of probiotics or probiotics and prebiotics in a randomized controlled trial. *American Journal of Nutrition*, 2008, 87, 1365-73.

17. Bettler J., Mitchel D.K., Kullen M.J. Administration of Bifidobacterium lactis with fructo-oligosaccharide to toddlers is safe and results in transient colonization. *International Journal of Probiotics and Prebiotics*, 2006, 1, 3/4 193-202.

18. Fisberg M., Maulen-Radovan I., Torno R., Crascoco M.T., Giner C.P., Martin F., Belinchon P., Costa C., Perez M., Caro J., Garibay E., Aranda J., Po I., Silva Guerra A., Martinez S., McCue M., Alarcon P., Corner G. Effect of oral nutrition supplementation with or without synbiotics on sickness and catch-up growth in preschool children. *International Pediatrics*, 2002, 17 (4), 216-22.

19. http://www.Müller.co.uk/a/pdf/eat-well/pre_and_probiotic_digestive_health.pdf

20. Molin G. Probiotics: compensating for a systematic error in modern diets. *Journal of Food Science and Technology*, 2007, 21 (4),17-21.

21. Guarner F. Prebiotics in inflammatory bowel disease, in *Handbook of Prebiotics*, Eds, Gibson G., Robertson M.B, Florida, CRC Press Boca Raton, 2008, 375-92.

22. Guandalini S. M. Probiotics for children: use in diarrhoea. *Journal of Clinical Gastroenterology*, 2006, 40 (3), 244-8.

23. Gianchetti P., Rizello F., Venturi A. Oral bacterio-therapy as maintenance treatment in patients with chronic pouchitis: a double-blind placebo controlled trial. *Gastro-enterology*, 2000, 119, 305-9.

24. Furrie E., MacFarlane S., Kennedy A., Cummings J.H., Walsh S.V., O'Neil D.A., MacFarlane G.T. Synbiotic therapy (*Bifidobacterium longum* / Synergy 1) initiates resolution of inflammation in patients with active ulcerative colitis: a randomised controlled pilot trial. *Gut*, 2005, 54, 242-9.

25. Klaenhammer T. R. Probiotics and Prebiotics, in *Food Microbiology –fundamental and frontiers 3rd Ed.*, Eds, Doyle D.P, Beuchat L. R, ASM Press, Washington D.C, USA, 2007, 891-910.

26. Ouwehand A., Kirjavainen P., Shortt C., Salminen S. Probiotics: mechanisms and established effects. *International Dairy Journal*, 1999, 9, 43-52.

27. Famularo G., Moretti S., Marcellini S., De Simone C. Stimulation of immunity by probiotics, in *Probiotics 2 Applications and Practical Aspects*. Ed. Fuller R. Chapman and Hall, London 1997, 133-61.

28. Devereux G. The immune system an overview, in *Nutrition and Immune Function*, Eds, Calder P., Field C., Gill H. Wallingford, UK, CABI Publishing, 2002, 1-20.

29. Guarner F., Malagelada J. Gut flora in health and disease. *The Lancet*, 2003, 361, 512-19 .

30. Forchielli M.L., Walker W.A. The role of gut associated lymphoid tissues and mucosal defence. *British Journal of Nutrition*, 2005, 93 Suppl, S41-8.

31. Kanamori Y., Sugiyama M., Komura M., Nakahara S., Sato K., Iwanka T., Yuki N., Morotomi M., Takahashi T., Tanaka R. Synbiotic Therapy: An important supportive therapy for pediatric patients with severe respiratory diseases. *International Journal of Probiotics and Prebiotics*, 2006, 19 (3), 161-8.

32. Hosano A., Kitazawa H. Yamaguchi T. Antimutagenic and antitumour activities of lactic acid bacteria, in *Probiotics 2 Applications and Practical Aspects*, Ed. Fuller R. London, Chapman & Hall, 1997, 89-132.

33. Gill H.S., Cross M.. Probiotics and immune function, in *Nutrition and Immune Function*. Eds, Calder P., Field C., Gill H.S. Wallingford, CABI Publishing, 2002, 251-72.

34. Ouwehand A., Tihonen K., Makivuokko H., Rautonen N. Synbiotics: combining the benefits of pre- and probiotics, in *Functional Dairy Products Vol 2*, Ed. Sarella M., Cambridge, England, Woodhead Publishing, 2007, 195-213.

35. www.worldhealthorganisation.com.

36. Decker K. Prebiotics and Probiotics: Banking on Synergism. Aug 2008. www.foodproductdesign.com

37. Van Loo J., Clune Y., Bennett M., Collins J. The SYNCAN project: goals, set-up, first results and settings of the human intervention study. *British Journal of Nutrition*, 2005, 93, S91-S8.

38. Roller M., Clune M., Collins K., Rechkemmer G., Watzl B. Consumption of prebiotic inulin enriched with oligofructose in combination with the probiotics *Lactobacillus rhamnosus* and *Bififobacterium lactis* has minor effects on selected immune parameters in polypectomised and colon cancer. *British Journal of Nutrition*, 2007, 97, 676-84.

39. Bartosch S., Woodmansey E.J., Paterson J.C.M., McMurdo M.E.T., MacFarlane G. Microbiological effects of consuming a synbiotic formula containing *Bifidobacterium bifidum*, *Bidobacterium lactis* and oligo-fructose in elderly persons, determined by real-time polymerase chain reaction and counting viable bacteria. *Clinical Infectious Diseases*, 2005, 40 (1), 28-37.

40. Bunout D., Barrera G., Hirsh S., Gattas V., de la Maza M., Haschke F., Steenhout P., Klassen P., Hager C., Avendano M., Patermann M., Munoz C. Effects of a nutritional supplement on the immune response of Cytokine production in free living Chilean elderly. *Journal of Parenteral and Enteral Nutrition*, 2004, 28, (5), 347-54.

41. Hamilton-Miller J., Smith C.T. Probiotic remedies are not what they seem. *British Medical Journal*, 1996, 312, 55-6.

42. Metchnikoff E. The prolongation of life. Heinemann. London, 1907.

43. Matilla-Sandhom T., Saarela M., de Vos W. Future developments of Probiotic dairy products, in *Probiotic Dairy Products*, Ed. Tamime A., Oxford, Blackwell Publishing, 2005, 195-207.

44. Decker K. The science behind the success of probiotics and prebiotics. *Food Products Design*, 2008, 18 (6) 76-85.

45. www.yoplus.com/whatisit.aspx

46. Heller L. Digestive health leads functional product launches, Sep 2008, www.foodanddrinkeurope.com

47. Minelli E.B, Benini A., Marzotto M, Ruzzenente O., Ferrario R. Assessment of novel probiotic *Lactobacillus casei* strains for the production of a functional dairy food. *International Dairy Journal*, 2004, 14, 723-36.

48. Lee Y.K. Introduction, in *Handbook of Probiotics*. Eds Lee Y.K, Nomoto K, Salminen S., Gorbach S.L. Chichester, UK, John Wiley & Sons, 1999, 1-12.

49. Huebner J., Weling R.L., Parkhurst A., Hutkins R.W. Effects of processing conditions on the prebiotic activity of commercial prebiotics. *International Dairy Journal*, 2008, 18, 287-93.

50. Cardarelli H.R., Aragon- Alegro J.H., Alegro J.H.A., de Castro I.A., Saad S.M. Effect of inulin and *Lactobacillus paracasei* on sensory and instrumental textural properties of functional chocolate mousse. *Journal of Science and Agriculture*, 2008, 88, 1318 -24.

51. Huebner J., Wehling R.L., Parkhurst A., Hutkins R.W. Effect of processing conditions on the prebiotic activity of commercial prebiotics. *International Dairy Journal*, 2008, 287-93.

52. Flambard B., Johansen E. Developing a functional dairy product: from research on *Lactobacillus helveticus* to industrial application of Cardi-o4 in novel antihypertensive drinking yogurts, in *Functional Dairy Products Vol 2*, Ed, Sarella M. Cambridge, England, Woodhead Publishing, 2007, 195-213.

53. Sveje M. Prebiotics and probiotics – improving consumer health through food consumption. *Nutracos*, 2007, 6, 28-31.

4. LEGISLATION

In common with other so-called functional foods, the future for products containing pre- and probiotics depends, to a significant extent, on the regulatory framework within which they are to be marketed.

4.1   European Community legislation

The addition of pre- and probiotics to foods is intended to be for the benefit of the population. Foods containing these would, therefore, fall under the general group of functional foods. There is no formal definition of 'functional food' at EC level, although it is understood that the term encompasses day-to-day foods eaten as part of a normal diet and not food supplements.

There is no positive list of compounds that are acceptable for use as prebiotics or probiotics; in effect, regulation is on a case-by-case basis. Examples of prebiotic compounds include fructooligosaccharides, galactooligosaccharides or lactulose, multifunctional oligosaccharides or lactosucrose. Such compounds can act in the gut to give the best conditions for the effectiveness of the particular food in question. Some compounds considered as prebiotics may have been used previously in foods, for example certain oligosaccharides have been used in the UK to increase the fibre content of food products. However, other compounds may not have been consumed to a significant degree within the Community prior to 15 May 1997 and may, therefore, fall under the scope of the Novel Foods Regulation (EC) No. 258/97 (as amended) (1). Genetically Modified (GM) foods originally fell under the scope of the novel foods Regulation,but in 2003, a new European legal framework for GM foods was adopted for their approval, labelling and traceability. GM foods were therefore removed from the scope of the novel food regulation which was amended by Regulation (EC) No. 1829/2003 on genetically modified food and feed (1).

4.2   Novel foods regulation

Before 1997, the individual approval procedures for novel foods and ingredients in each Member State were proving a significant barrier to free trade within the European Union. It was therefore deemed appropriate that EC legislation should be developed in order to try to reduce the burden of approvals, to one acceptable to all Member States. The purpose of this Regulation is to ensure that the free movement of novel foods is not hindered throughout the Community, while protecting the interests of consumers, especially in respect of safety, health and information. In the same way as the regulatory framework for functional foods is being assessed, the Novel Foods Regulation is expected to provide the basic regulatory structure for the development of food innovation.

There are four novel food categories:

1. Foods and food ingredients with a new or intentionally modified primary molecular structure; e.g.: Salatrim (structured chain of fatty acids used as reduced energy replacement for fats and oils), D-tagatose (similar to fructose with 75% sweetness of sucrose, used as a source of carbohydrates, non cariogenic and probiotic properties).

2. Foods and food ingredients consisting of or isolated from micro-organisms, fungi or algae; e.g.: Fungal lycopene from Blakeslea trispora, DHA-rich oil from microalgae Schizochytrium sp.

3. Foods and food ingredients consisting of or isolated from plants and food ingredients isolated from animals, except for foods and food ingredients obtained by traditional propagating or breeding practices and which have a history of safe food use; e.g.: noni juice, tomato lycopene, plant sterols, baobab dried fruit pulp.

4. Foods and food ingredients to which a production process not currently used has been applied, where such a process gives rise to significant changes in the composition or structure of the foods or food ingredients that affect their nutritional value, metabolism or level of undesirable substances. e.g.: Trehalose produced with new enzymic process, nanomaterials.

Each of the 27 EU Member States has its competent authority on Novel Foods. They ultimately make the decision as to whether an ingredient should be considered novel or not in the EU. Where necessary, they may discuss the status of a novel food during Working Group meetings chaired by the Commission in Brussels or it may be decided that a decision needs to be taken between Member States by voting at the Standing Committee on Food Chain and Animal Health whether a type of food or food ingredient falls under the scope of this Regulation.

From these discussions between Member States, the Commission has compiled a Novel Food catalogue listing foods and food ingredients that are considered novel or not, which is publicly available on the Commission's website.

There are two different procedures required in order to place novel foods or food ingredients on the EU market:

The full novel food procedure which applies to the four novel food categories mentioned above. This involves submitting a full dossier demonstrating the safety of the novel product, including data on:

• specification of novel food

• effect of production process

• history of consumption of source

- anticipated intake and extent of use

- information on previous human exposure

- nutritional information

- microbiological information

- toxicological/allergenicity information

- labelling information.

It can take between 8 months to 2 years to obtain an authorisation to market a novel product under this procedure.

The simplified, or notification procedure is based on substantial equivalence of a novel food with an existing food. It is only applied foods or food ingredients falling under the following two out of the four novel food categories mentioned above:

- Categoriy No.2: foods and food ingredients consisting of, or isolated from micro-organisms, fungi or algae,

- Category No. 3:. foods and food ingredients consisting of, or isolated from plants and food ingredients isolated from animals, except for foods and food ingredients obtained by traditional propagating or breeding practices, and which have a history of safe food use.

This simplified, or notification procedure involves submitting a dossier containing scientific evidence available and generally recognised, or on the basis of a positive opinion delivered by one of the 27 EU competent authorities on a dossier, showing the novel food is substantially equivalent to existing foods or food ingredients as regards to the following 5 criteria:

- composition,

- nutritional value,

- metabolism,

- intended use and

- the level of undesirable substances.

Where necessary, it may be determined by the EU Member States via the Standing Committee on Food Chain and Animal Health whether a type of food or food ingredient can be considered substantially equivalent to an existing product.

The full novel food procedure would generally be applied for authorising novel microorganisms such as bifidobacteria, *Lactobacillus* and possibly other miscellaneous probiotics such as lactococci or yeasts. The European Commission publishes and updates the list of full novel food applications submitted under (EC) 258/97 (as amended). Until now, there are no novel food applications that have been made for the authorisation of a pre- or probiotic micro-organism as a novel food ingredient. Each Member State has a nominated competent authority to which evidence can be submitted in order to obtain an opinion on whether an ingredient is novel or not. The appropriate body in the UK is the Food Standards Agency (Novel Food Unit).

4.2.1   Labelling of novel ingredients

Without prejudice to other requirements of Community law concerning the labelling of foodstuffs, the following additional specific labelling requirements apply, to ensure that the consumer is informed of:

    (i)   any characteristic or food property such as:

- composition;

- nutritional value or nutritional effects;

- intended use of the food,

that renders a novel food or food ingredient no longer equivalent to an equivalent existing food or food ingredient. In this case, the labelling must mention the characteristics or properties modified, accompanied by an indication of the method by which that characteristic or property was obtained;

    (ii)   the presence in the novel food or food ingredient of material that is not present in an existing equivalent foodstuff, and which may have implications for the health of certain sections of the population;

    (iii)   the presence in the novel food or food ingredient of material that is not present in an existing equivalent foodstuff, and which may give rise to ethical concerns. In the absence of an existing equivalent food or food ingredient, appropriate provisions shall be adopted where necessary to ensure that consumers are adequately informed of the nature of the food or food ingredient.

4.3   Labelling issues

There are no specific labelling rules for pre- and probiotics. In the context of labelling functional foods, it is the prospect of a claim and also the type of claim given on the label that will inform the consumer about the nature of the product. Nutrition and health claims are controlled at a European level. General principles on misleading labelling must also apply as in Commission Directive 2000/13/EC (2). EC legislation prohibits the use of medicinal claims in food labelling, i.e. claims that a food can cure, treat or prevent human disease.

4.4   Nutrition and health claims

The final legislation on nutrition and health claims was published in its correct format on 18 January 2007. Regulation (EC) No. 1924/2006 (3) aims to provide a high level of protection for human health, and to promote the protection of consumer interests by ensuring foods carrying nutrition, and health claims are labelled and advertised in a clear manner enabling informed choice by consumers.

Nutrition claims are restricted to those listed in the Annex to the Regulation, which gives indicative wording and sets conditions that a product must meet to carry the claim. Health claims are divided into two classes; the first (Article 13 claims) encompasses nutrient function claims, references to psychological and behavioural functions, slimming and satiety. A register of such claims will be set up by 31 January 2010. The Regulation regards reduction of disease risk claims and those referring to children's development and health differently, with a more detailed process to achieve authorisation for use on foods. To ensure harmonised scientific assessment, the European Food Safety Authority (EFSA) will carry out assessments of proposed health claims. Timescales are set within the Regulation for the authorisation procedure; comments may be submitted after the EFSA has given its opinion. Claims relating to the prevention and treatment of disease will still remain prohibited (as stated in Directive 2000/13/EC on general labelling of food).

According to Community Guidance on the implementation of the Regulation, a claim such as 'contains prebiotics/probiotics' shall be treated as a health claim and not a nutrition claim even if 'contains' claims normally are regarded as nutrition claims. This is due to the fact that the term prebiotics/prebiotics refers to a group of substances with a specific functional effect. Hence, the claim 'contains prebiotics' is a health claim and must be authorised accordingly. However, if a 'contains claim' refers to a specific prebiotic fibres or a specific bacteria, for example 'contains *Lactobacillus* ', the claim is to be treated as a nutrition claim. Hence, for pre- or probiotic claims to be able to be used, they have to be included in the EC register of approved health claims. At the time of publication of this book, this register had not yet been finalised and no comment can be made on the claims on the list.

The Regulation allows for restrictions on the use of claims linked to a foodstuff's nutrient profile.

4.5   EU Member States

If a pre- or probiotic is submitted to a competent authority for a decision on whether or not it is a novel food, the procedures as established in the Novel Foods Regulation EC (No) 258/97 will be applicable. All EU Member states also have to comply with Regulation (EC) 1924/2006 on nutrition and health claims. However, individual Member States may have additional guidance relating to pre- and prebiotics that is applicable in their own countries.

'Fibre' has recently been defined by Directive 2008/100/EC:

"fibre" means carbohydrate polymers with three or more monomeric units, which are neither digested nor absorbed in the human small intestine and belong to the following categories:
— edible carbohydrate polymers naturally occurring in the food as consumed;
— edible carbohydrate polymers that have been obtained from food raw material by physical, enzymic or chemical means, and which have a beneficial physiological effect demonstrated by generally accepted scientific evidence;
— edible synthetic carbohydrate polymers that have a beneficial physiological effect demonstrated by generally accepted scientific evidence.'

Any nutrition or health claims based on fibre would need to use this definition, although the definition will not be legally enforceable until 31st October 2012, after it is transcribed into national law by 31st October 2009 at the latest.

This new definition may class ingredients such as oligofructose, and other extracted or synthetic substances as fibre provided they have 'a beneficial physiological effect demonstrated by generally accepted scientific evidence'. This will presumably be clarified from a health claim perspective when the Article 13 claims list is published.

4.5.1   France

Probiotic cultures including *Lactobacillus* and *Bifidus* sub-species are authorised for use in yoghurts and fermented milks. In particular for yoghurts, the bacteria are *Streptococcus thermophilus*, and *Lactobacillus delbrueckii* subsp *bulgaricus*. Bacterial cultures are also permitted in foods for particular nutritional uses, in particular in the Order of 4 August 1986, as amended. A one million /gram minimum for the live bacteria applies.

The French Food Health and Safety Agency (AFSSA) issued a favourable Opinion on 22 December 2000 for claims relating to the effects of inulin on human intestinal flora. The Opinion acknowledged that the consumption of inulin with an average degree of polymerisation of 9 units (DP9) increased the concentration of bifidobacteria in the intestine, but judged that there was not enough data to evaluate the effects of inulin on their composition and metabolic activity. Therefore, to claim "the oral consumption of DP9 inulin significantly increases the population of Bifidobacteria in the human intestine" was deemed acceptable. This claim cannot be accompanied by any claim of a preventative beneficial effect or curative effect on the health. The claim is to be accompanied by the following statements:

-    minimum consumption of 9g/day to obtain the claimed effect

-    possible incidence of intestinal upset (or problems) if more than 20g/day consumed

The AFSSA also noted that inulin can pose a risk for those with allergies and recommended that the labelling of final products with inulin ingredients should mention its presence.

4.5.2   UK

In the UK, increased recognition of the role of the diet in maintaining good health and an anticipated growth in the functional foods market have emphasised the need to prevent the use of false, exaggerated, misleading and prohibited health claims in order to assist fair trade and to protect the consumer from false information.

The Nutrition and Health Claims Regulation, 1924/2006, is enforced in England under the Nutrition and Health Claims (England) Regulations 2007 as amended (SI 2007 No. 2080). Generic health claims, including those relating to pre and pro-biotics, may be added to the Article 13 claims list in the future.

Both pre and pro-biotic claims were included in the UK claims list that is currently being assessed by the EFSA. Nutrition claims such as 'rich in fibre' are now covered by the EC Regulation, as are most other permitted Nutrition claims, although there is a transitional period for nutrition claims that have been approved in a member state and existed before 1st January 2006. Such claims, provided they are not claims detailed in the Annex of the EC Regulation will be permitted until 19th January 2010.

Historically, the Joint Health Claims Initiative (JHCI) was a collaborative body between enforcement, consumer and industry groups, with the aim of providing substantiation for health claims in the UK. The JHCI officially ceased operating in April 2007, although their Code of Practice on Health Claims is still of some use today. The code outlines the general principles to be applied when making a health claim, and contains details on the substantiation of health claims and examples of the type of health claim that may be considered acceptable.

The JHCI website and the information on it will be available until the transition period for the claims regulation ends in 2010.

4.5.3   Sweden

As a member of the European Union, Regulation (EC) No. 1924/2006 on nutrition and health claims should apply in Sweden. Before the EC provisions on nutrition and health claims came into force, a voluntary self-regulating programme on health claims used in the labelling and marketing of foods was in place (5). Participants in the programme included industry representatives, and trade bodies, agricultural producers and retailers' representatives. Under the agreement, eight connections between diet-related diseases, or risk factors for these, was regarded as well-founded and formed the basis for acceptable health claims for marketing purposes. Certain products in this area have been registered as natural remedies, i.e. are medicinal products that are suitable for home cures in accordance with well-proven Swedish traditions, or traditions in other countries considered to have similar medicinal traditions to those in Sweden. New products must be approved by the Medical Products Agency.

4.6   International Developments

For prebiotics and probiotics that may fall under the provisions of the Novel Foods Regulation, the practical implications of the approvals process still remain to be tested. Approval would not automatically mean that the ingredient or component in question had a health-related role in the diet; any such claims would require justification by appropriate scientific evidence. Outside Europe, the same dilemma is facing other regulatory authorities as, after all, the search for new, functional ingredients is a global one and questions of approval are mirrored the world over.

4.6.1   FAO/WHO

The Joint FAO/WHO Working Group on Drafting Guidelines for the Evaluation of Probiotics in Food released guidelines for the Evaluation of probiotics in Food on the 1 May 2002
(http://www.who.int/foodsafety/fs_management/en/probiotic_guidelines.pdf#s earch=%22fao%2Fwho%20probiotics%22).

The Guidelines report that, historically, *Lactobacilli* and *Bifidobacteria* associated with food have been considered to be safe (Adams & Marteau, 1995). Their occurrence as normal commensals of the mammalian flora and their established safe use in a diversity of foods and supplement products worldwide supports this conclusion.

Regarding species and strains, they state that, in the case of *S. thermophilus* and *L. delbrueckii* ssp. *bulgaricus* or in other cases where there is suitable scientific substantiation of health benefits that are not strain-specific, individual strain identity is not critical. Hence, for these cases, exemptions can possibly be made.

The Guidelines also recommend that the nomenclature of the bacteria should conform to the current, scientifically recognised names. Protracted use of older or misleading nomenclature is not acceptable on product labels. The use of incorrect names does not properly identify the probiotic bacterium in the product and forces consumers and regulatory agencies to make assumptions about the identity of the real bacterium being sold. The current nomenclature can be retrieved as follows:

An approved list of bacterial names (Int. J. Syst. Bacteriol, 1980, 3, 225-420), is available on http:/ijs.sgmjournals.org/cgi/reprint/30/1/225.

4.6.2   Japan

Japan, thought of as the 'home' of the functional foods concept, considers functional foods as 'foods for specified health use' and defines them as foods that, based on the knowledge concerning the relationship between foods or food components and health, are expected to have certain health benefits and to have been licensed to bear labelling claiming that a person using them may expect to obtain that health benefit through the consumption of these foods. To be labelled as suitable for a specified health use, foods must go through the approval procedure.

Among the wide range of products approved as foods for specified health use are those shown in Table 4.II.

TABLE 4.II
Some approved health claims in Japan

Food type	Functional component	Health Claims licensed by Ministry of Health and Welfare
Lactic acid bacteria drink	Xylo-oligosaccharide	A drink maintaining a healthy balance of intestinal bifidobacteria that improves and maintains the condition of the intestine.
Table sugar	Fructo-oligosaccharide	The food increases the number of bifidobacteria and maintains the good condition of the intestine.
Carbonated beverage	Soya-bean oligosaccharides	The product increases the rate of bifidobacteria growth in the intestine and maintains good intestine condition. It is suitable for improving bowel condition.

Products such as 'Yakult' were first marketed in Japan and have now extended to an international market.

4.6.3 Australia

In Australia and also in New Zealand, Standard 1.5.1 of the Australia and New Zealand Food Standards Code regulates the sale of novel food and novel food ingredients. The standard prohibits the sale of these foods unless they are listed in the Table included in this standard. Novel foods are considered as a subset of non-traditional foods, i.e. foods that do not have a history of significant human consumption by the broad community in Australia and New Zealand. Novel foods are classified as likely to fall into one of four areas, one of which is microorganisms and another is dietary macrocomponents. Probiotics would be covered by the former group, and prebiotics possibly by the latter. The types of microorganism suggested as being potentially covered by novel foods provisions include *Lactobacillus* strains and *Bifidobacterium* strains. Concerns brought forward in respect of these novel microorganisms include the potential effect of colonisation of the gastrointestinal tract and the potential for these organisms to affect the absorption of nutrients and other biologically active substances.

4.6.4   USA

There is not currently a legal definition for "probiotics" in the USA. The Food and Drug Administration's (FDA) Draft *"Guidance for industry on complementary and alternative medicine products and their regulation by the food and drug administration"*, December 2006 notes the following:

- "Probiotics" have been defined as live microbial food supplements that beneficially affect the host animal by improving its intestinal microbial balance (this also cites FAO/WHO definition)

- "For purposes of this document, we will consider probiotics to refer to whole, live microorganisms that are ingested with the intention of providing a health benefit (such as supporting digestion and nutrient adsorption in the intestine)".

The Draft notes further that:

Probiotics may be regulated as dietary supplements, foods, or drugs under the Act (the Federal Food, Drug and Cosmetic Act), depending on the product's intended use. Other factors may also affect the classification of the product, e.g., whether the product contains a "dietary ingredient" as defined in section 201(ff)(l) of the Act, whether it is represented as a conventional food or as a meal replacement, and, for probiotics used as ingredients in a conventional food, whether the ingredient is generally recognised as safe for its intended use.

The Draft Guidance defines prebiotics as: "Prebiotics have been defined as nondigestible food ingredients that beneficially affect the host by selectively stimulating the growth and/or activity of one or a limited number of bacteria in the colon".

As there is not a specific legal control of the use of pro- or prebiotics, the following general requirements should be considered.

In the USA, substances intended for use in the manufacture of foodstuffs for human consumption are classified into three categories:

- food additives, defined as substances the intended use of which results or may reasonably be expected to result, directly or indirectly, either in their becoming a component of food or otherwise affecting the characteristics of food;

- prior-sanctioned food ingredients, which are substances that received official approval for their use in food by the FDA or the US Department of Agriculture (USDA) prior to 1958;

- substances generally recognised as safe (GRAS), which are substances that are recognised by the experts experienced and trained in evaluating the safety of food substances to be safe for their intended use in food.

When Congress established the definition of a food additive, they recognised that many substances that are intentionally added to food would not require a formal review of their safety, and as a result provision was made for prior-sanctioned food ingredients and GRAS substances.

The lists of GRAS substances in the regulations are not exhaustive and the regulations state that it would be impractical to list all substances that are GRAS for their intended use.

A substance that is GRAS for a particular use may be marketed for that use without agency review and approval, provided that the manufacturer can prove that the use of the substance is GRAS, as determined by qualified experts. Therefore, a manufacturer may independently determine that the use of a substance is GRAS. The determination requires both technical evidence of safety, and a basis to conclude that the technical evidence is generally known and accepted.

Manufacturers may petition the FDA to affirm that a substance is GRAS under certain conditions of use, as a voluntary process to provide for official recognition of lawfully made GRAS determinations. As well as those substances that are listed or affirmed in the regulations as GRAS, there are also some substances that have been 'notified' as GRAS, and do not appear in the regulations.

There have been GRAS notifications submitted for two strains of bacterial cultures, and for several ingredients that have prebiotic properties, including polydextrose, Isomaltulose, Fructo-oligosaccharides, d-Tagatose.

4.6.5   *Codex Alimentarius*

At Codex level, there are no specific provisions or definitions for pre- and probiotics. Although discussions are taking place on health claims in general, these are at an early stage (Step 3), with interest focusing on the potential need for such claims. Health claims have been subdivided into two types, namely enhanced function claims and reduction of disease risk claims (5). As the former concern specific beneficial effects of foods and their constituents on biological activities or physiological functions, and relate to a positive contribution to health, claims regarding the efficacy of probiotics would fall under this category.

4.7   Summary

How far can manufacturers legitimately go in making claims concerning the potential health benefits of their product, and what scientific evidence is available to support these claims? There must be sufficient flexibility in the law to allow manufacturers to attract consumers by means of the information on the label, while making sure that such labelling is within the boundaries set by the law. If functional foods such as foods containing pre- and probiotics are to have a positive future in the world market, consumers must have the confidence to believe that the claims that adorn their labels can be justified on the basis of sound scientific evidence.

4.8   References

1. Official Journal of the European Communities, 40 (L43), 14/2/97, pp 1-7. See amended consolidated version at: http://eur-lex.europa.eu/LexUriServ/LexUriServ. do?uri=CONSLEG:1997R0258:20040418:EN:PDF

2. Official Journal of the European Communities, (L109), 6/5/2000, pp 29-56.

3. Official Journal of the European Communities

4. Health Claims in the labelling and marketing of food products – The Food Sector's Code of Practice, September 2004.

5. Proposed draft recommendations for the use of health claims, alinorm 99/22A, Appendix VII.

5. SUPPLIERS

A selection of some of the leading suppliers of probiotics and prebiotics is presented in this section. A summary table showing the prebiotics supplied by each company is presented on page 148.

PREBIOTICS

BENEO – Orafti
Aandorenstraat 1
B-3300 Tienen,
Belgium
Tel: + 32 16 801 301
Fax: + 32 16 801 308
Supplies: inulin, oligofructose

Beghin Meiji Industries
14 Bd du General Leclerc
F-92572 Neuilly sur Seine Cedex
France
Tel: +33 14 14 31 148
Fax: +33 14 14 31 302
Supplies: fructooligosaccharides
(Actilight)

Calpis Co.
2-20-3 EBISHU-Nishi
Shibuya-ku
Tokyo 150
Japan
Tel: +81 3 3463 2111
Fax:+81 3 3770 5374
Supplies: soybean oligosaccharides

Chephasaar Chem.- pharm. Fabrik GmbH
Muehlstrasse 50
D-66386 St. Ingbert
Germany
Tel: +49 6894 9710
Fax: +49 6894 971199
Supplies: lactulose

CoSucra
Rue de la Sucrerie, 1
B-7740 Warcoing
Belgium
Tel: +32 69 44 66 00
Fax: +32 69 44 66 22
Supplies: inulin,
fructooligosaccharides and
oligofructose

Danipharm APS
Englandsvej 350-6
DK 2770 Kastrup
Denmark
Tel: +45 70 10 3020
Fax: +45 32 50 1605
Supplies: lactulose

DKSH (UK)
3rd Wellington House
60-68 Wimbledon Hill Road
London, SW19 7PA
UK
Tel: +44 20 8879 5500
Fax: +44 20 8879 5501
Supplies: inulin,
fructooligosaccharides, oligofructose,
lactulose (Raftiline)

Ensuiko Sugar
13-46 Daikoku-cho
Tsurumi-ku
Yokohama
Japan 230
Tel: +81 45 501 1251
Fax: +81 45 501 1257
Supplies: lactosucrose

Friesland Foods
Corporate Research
PO Box 87
NL-7400 AB Deventer,
The Netherlands
Tel: + 31 570 695 998
Fax: +31 570 695 918
Supplies: galactooligosaccharides

Hayashibara
1-2-3 Shimoishii
Okayama-shi
Okayama-ken
Japan 700 0907
Tel: +81 86 224 4311
Fax: +81 86 233 2265
Supplies: isomaltooligosaccharides

Meiji Seika Kaisha
2-4-16 Kyobashi
Chuo-ku
Tokyo 104
Japan
Tel: +81 3 3272 6511
Fax: +81 3 3271 3528
Supplies: fructooligosaccharides

Milei GmbH
Rosensteinstr. 22
70191
Stuttgart
Germany
Tel: +49 711 981 780
Fax: +49 711 981 7825
Supplies: lactulose

Moringa Milk Industry
5-33-1 Shiba
Minato-ku
Toyko 108
Japan
Tel: +81 3 3798 0111
Fax: +81 3 3798 0101
Supplies: lactulose

Nihon Shokuhin Kako
5-33-8 Chidadaya
Shibuya-ku
Toyko
Japan 151
Tel: +81 3 5360 4411
Supplies: isomaltooligosaccharides,
glucooligosaccharides

Nissan Sugar
14-1 Nihonbashi-Koamicho
Chuo-ku
Tokyo
Japan 103
Tel: +81 3 3668 2422
Fax: +81 3 3668 1126
Supplies: Galactooligosaccharides

Sensus Operations BV
Oosteliijke Havendijk 15
4704 RA Roosendaal
Netherlands
Tel: +31 165 582 577
Fax: +31 165 567 796
Email: Info.Sensus@Sensus.nl
Supplies: inulin,
fructooligosaccharides, oligofructose

Showa Sangyo
2-2-1- Uchi Kanda
Chiyoda- ku
Tokyo
Japan 101
Tel: +81 3 3257 2011
Fax: +81 3 3257 2097
Supplies: isomaltooligosaccharides

Snow Brand Milk Products
13 Honshio-cho
Shinjuku-ku
Tokyo 160
Japan
Tel:+81 3 3226 2111
Fax:+81 3 3226 2150
Supplies: galactooligosaccharides

Solvay Pharmaceuticals
PO Box 900
C.J. Van Houtenlaan 36
NL – 1381 CP Weesp
The Netherlands
Tel: +31 294 47 7000
Fax: +31 294 48 0253
Supplies: galactofructose

Suntory
2-1-40 Dojimahama
Kita-ku
Osaka 530-8203
Japan
Tel: +81 6 346 1131
Fax: +81 6 345 1169
Supplies: xylooligosaccharides

Syral
Z.I et Portuaire
BP 32
Marckolsheim F-67 390
France
Tel: +33 388 58 6060
Fax: +33 388 58 6061
Supplies: fructooligosaccharides

Yakult Honsha
1-1-19 Higashi Shimbashi
Minato-ku
Toyko 105
Japan
Tel: +81 3 3574 8960
Fax: +81 3 3575 1636
Supplies: galactooligosaccharides

PROBIOTICS

Biogaia Biologics AB
PO Box 3242
S-103 64 Stockholm
Sweden
Tel: + 46 8 555 293
Fax: +46 8 555 293 01
Supplies: exclusive rights to use
Lactobacillus reuteri in products.

Cargill Texturising Solutions
15407 McGinty Rd
46 Wayzata MN 55391
USA
Tel: +1 877 765 8867
Fax: +1 952 742 1087

Cell Biotech
134 Gaekok-RI
Wolgot-Myun
Gimpso-Si
Gyuunggi-Do
Korea
Tel: +82 31 987 6205
Fax: +82 31 987 6209

Chr Hansen A/S
Boge Alle 10-12
2970 Horsholm
Denmark
Tel: +45 45 767676
Fax: +45 45 760848

Cognis -GmBH & Cohs
Robert-Hansen-Str 1
D-89257 Illertissen,
Germany
Tel: +49 730 313 000
Fax: +49 730 313 210

Danisco
Edwin Rahrs Vej 38
DK-8220 Brabrand,
Denmark
Tel: +45 8984 3500
Fax: +45 2948 4435
Supplies: Bifidobacterium lactis
HN019, Lactobacillus acidophilus
NCFM® (HOWARTY®)

Degussa Bioactives
Freising
Germany
Tel: +49 2159 697061
Fax: +49 2159 697060

DSM Food Specialties
PO Box 1
NL-2600 MA Delft,
The Netherlands
Tel: +31 152 793 3474
Fax: +31 152 794 3540
Supplies: Lactobacillus casei
(LACTI L26)

Fortitech Ltd
Egegaardsvej 9
DK-4621 Gadstrup,
Denmark
Tel: +45 5824 0500
Fax: +45 5824 0580

Genibo
Voie Haussman
Z.I du Couseran
09190 Lorp-Sentaraille
France
Tel: +33 561 048 142
Fax: +33 561 048 060

Giellepi Chemical Spa
Via G. Verdi, 41/Q
20038 Seregno
Milan
Italy
Tel: +39 0362 240116
Fax: +39 0327806

Greentech
Biopole Clermont
Limagne 63360
Saint Beauzire
France
Tel: +33 473 33 9900
Fax: +33 473 339131

Institut Rosell – Lallemand
8480 Saint-Laurent Boulevard
Montreal, Quebec H2P 2M6
Canada
Tel: +1 514 381 5631
Fax: +1 514 383 4493

J. Rettenmaier & Söhne
Holzmuhle 1
73494 Rosenburg
Germany
Tel: +49 7967 15200
Fax: +49 7967 15222

Probi AB
Solvegatan 41
SE-223 70 Lund,
Sweden
Tel: +46 462 868 920
Fax: +46 462 868 928
Supplies: exclusive rights to use
Lactobacillus plantarum 299, 271

Probiotical Srl
Via E Mattei 3,
28100 Novara
Italy
Tel: +30 0321 465 933
Fax: +39 0321 492 693

SKW Biosystems
4 Places des Ailes
F 92641 Boulogne Billancourt
Cedex
France
Tel: +33 1 47 12 25 58
Fax:+33 1 47 12 27 70

UAS Laboratories
9953 Valleyview Rd
Eden Prairie
MN 55344
USA
Tel: +1 952 935 1707
Fax: +1 952 935 1650
Supplies: Bifidobacteria and
Lactobacillus acidophilus cultures
under the BIOghurt and BIOgarde
trademarks

Prebiotic Suppliers

Supplier	Inulin	Fructo-oligo-saccharides	Oligo-fructose	Galacto-oligo-saccharides	Soya-bean oligo-saccharides	Isomalt oligo-saccharides	Gluco-oligo-saccharides	Xylo-oligo-saccharides	Galacto-fructose	Lacto-sucrose
Beneo Orafti	?	?	?						?	
Beghin Meiji		?								
Calpis Co.					?					
Chephasaar									?	
Cosucra	?	?	?							
Danipharm										
Ensuiko Sugar									?	?
Friesland Foods						?				
Garuda International	?									
Hayashibara		?				?				
Meiji Seika Kaisha										
Milei									?	
Moringa Milk									?	
Nihon Shokuhin Kako						?	?			
Nissan Sugar				?						
Resolution									?	
Sensus	?	?	?							
Showa Sangyo						?				
Siber Hegner	?	?	?				?			
Snowbrand Milk				?						
Solvay									?	
Suntory								?		
Yakult Honsha				?						

148

INDEX